D1612179

ACS SYMPOSIUM SERIES **633**

Flavor–Food Interactions

Robert J. McGorrin, EDITOR
Kraft Foods, Inc.

Jane V. Leland, EDITOR
Kraft Foods, Inc.

Developed from a symposium sponsored
by the Division of Agricultural and Food Chemistry
at the 208th National Meeting
of the American Chemical Society,
Washington, DC,
August 21–25, 1994

American Chemical Society, Washington, DC

Library of Congress Cataloging-in-Publication Data

Flavor–food interactions: developed from a symposium sponsored by the Division of Agricultural and Food Chemistry at the 208th National Meeting of the American Chemical Society. Washington, DC. August 21–25, 1994 / Robert J. McGorrin, editor, Jane V. Leland, editor.

p. cm.—(ACS symposium series, ISSN 0097–6156: 633)

Includes bibliographical references and indexes.

ISBN 0–8412–3409–4

1. Flavor—Congresses.

I. McGorrin, Robert J., 1951– . II. Leland, Jane V. III. American Chemical Society. Division of Agricultural and Food Chemistry. IV. American Chemical Society. Meeting (208th: 1994: Washington, D.C.) V. Series.

TP372.5.F544 1996
664′.07—dc20

96–20197
CIP

This book is printed on acid-free, recycled paper.

PRINTED IN THE UNITED STATES OF AMERICA

Foreword

THE ACS SYMPOSIUM SERIES was first published in 1974 to provide a mechanism for publishing symposia quickly in book form. The purpose of this series is to publish comprehensive books developed from symposia, which are usually "snapshots in time" of the current research being done on a topic, plus some review material on the topic. For this reason, it is necessary that the papers be published as quickly as possible.

Before a symposium-based book is put under contract, the proposed table of contents is reviewed for appropriateness to the topic and for comprehensiveness of the collection. Some papers are excluded at this point, and others are added to round out the scope of the volume. In addition, a draft of each paper is peer-reviewed prior to final acceptance or rejection. This anonymous review process is supervised by the organizer(s) of the symposium, who become the editor(s) of the book. The authors then revise their papers according to the recommendations of both the reviewers and the editors, prepare camera-ready copy, and submit the final papers to the editors, who check that all necessary revisions have been made.

As a rule, only original research papers and original review papers are included in the volumes. Verbatim reproductions of previously published papers are not accepted.

ACS BOOKS DEPARTMENT

Contents

Preface

THE INTERACTION OF FLAVOR COMPOUNDS with food systems is an important factor, although not the only one, in overall flavor perception. Food-matrix components including proteins, carbohydrates, and "replacer" ingredients such as sugar- or fat-substitutes are capable of "binding", absorbing, entrapping, or encapsulating volatile flavor components. In lipid–water systems and emulsions, flavor compounds partition between the fat and water phases as a function of the physicochemical properties of the lipid and the flavor. All of these situations affect perceived flavor either by heightening or reducing the impact of individual components, and thus altering overall balance. The study of flavor–food matrix interactions is often extremely difficult because of the complex nature of food ingredients and the multiplicity of potential interactions with flavor constituents. Because of the complexity, many researchers (including a majority of authors in this book) choose to study model systems which have only binary or ternary interactions.

The symposium from which this book was developed is a first attempt to comprehensively discuss the topic of interactions between flavor and nonflavor components. Recent pragmatic interest in this subject has been tied to the development of low-fat and sugar-free foods and the corresponding flavor issues that resulted. Coverage in this book includes a compilation of the most significant recent work on interactions between volatile flavoring substances and food constituents. The symposium was international in scope, including presenters from Northern Ireland, Scotland, Great Britain, France, Germany, and the United States.

The chapters in this book are organized into four major sections. The first section provides an overview of flavor–food interactions and their effect on overall flavor perception. The second section focuses on interactions with major food components (lipids, proteins, starches), gelling agents, and emulsions. Traditional flavor encapsulation was beyond the scope of this book and is more thoroughly covered in the ACS Symposium Series' *Flavor Encapsulation* (No. 370) and *Encapsulation and Controlled Release of Food Ingredients* (No. 590). Although the major emphasis of the book is food–system interactions, three chapters discuss flavor interactions in chewing gum, a dentifrice matrix, and packaging materials. The last subject is more extensively discussed in the ACS Symposium Series' *Food and Packaging Interactions I* and *II* (Nos. 365 and 473). The third section presents applications of several measurement tools, including gas chromatography–olfactometry and principal component analysis. The final section discusses various approaches to the study of flavor interactions in complex food systems. This book is intended to serve as a reference for flavor chemists and food-product-development scientists who need to understand how flavors are influenced by various food components.

Acknowledgments

We thank those scientists who served as referees for the chapters in this book, and especially Susan Bodett for valuable assistance with figure graphics and in correcting and retyping many of the manuscripts. Finally, we acknowledge with sincere appreciation the financial support of our sponsors, Kraft Foods and DRAGOCO AG, without which the symposium would not have been possible.

ROBERT J. MCGORRIN
JANE V. LELAND
Kraft Foods, Inc.
801 Waukegan Road
Glenview, IL 60025

March 6, 1996

Introduction

Robert J. McGorrin

FLAVOR PERCEPTION IN FOODS IS HIGHLY INFLUENCED by interactions between volatile aroma compounds with a variety of non-flavor food matrix components. There are several hypotheses regarding the nature of these interactions, which can simplistically be classified into three types:

— Binding (retention or absorption of volatile compounds onto non-volatile substrates).
— Partitioning (distribution of flavor substances between the oil, water, and gas phases).
— Release (availability of flavor compounds from the bulk food into the gas phase for sensory perception)

The type of interaction depends on the physico-chemical properties of the flavorants and their relative concentrations. According to Maier (*1*), Solms et al. (*2*), and Voilley et al. (*3*), the fixation of aroma substances in food results from several processes:

— Covalent, irreversible bonding (e.g., aldehyde or ketone fixation by protein amino groups).
— Hydrogen bonding (e.g., interactions between polar, volatile alcohols and hetero atoms (N, S, O) of food substrates)
— Hydrophobic bonding (weak, reversible Van der Waals interactions between apolar volatile flavor compounds and fat molecules)
— Formation of inclusion complexes (e.g. volatile flavor complexes with β-cyclodextrin).

As food is consumed, it is masticated, diluted with saliva, and warmed to body temperature (37 °C). Volatile flavor compounds that are released during this process diffuse through the retronasal cavity towards the olfactory epithelium, where the flavorant is perceived (*4*). The relative balance of different flavor-ingredient combinations ultimately influences the overall flavor perception. Understanding the processes which influence the release and binding behavior of flavor volatiles from the food matrix is of major significance for improving flavor quality.

Interactions with Proteins and Amino Acids

The most frequently studied flavor ingredient interaction, as reported in the literature, is the binding of volatiles to proteins, especially soy protein (*5*). From binding studies, it was demonstrated that heat denaturation of the protein increased the binding capacity of aliphatic aldehydes and ketones; for alcohols there are contradictory results. The effects of protein binding can be rather complicated, since the binding of volatiles is strongly dependent on the concentration and water content of the protein. Anhydrous zein proteins have been shown to bind propanol (*6*). Sodium caseinate binds a higher degree of ketones and esters than starch (*3*). For most cases, the binding of flavoring substances to proteins depends on the degree of denaturation, the temperature and the pH. In general, hydrocarbons, alcohols, and ketones are reversibly bound to proteins through hydrophobic interactions and hydrogen bonding. However, aldehydes tend to react chemically with protein amino groups, resulting in irreversible binding (*6*).

Free amino acids can bind with a series of volatile flavoring substances in aqueous media. Ketones and alcohols are reversibly bound via hydrogen bonding to the amino or carboxyl groups of amino acids, while as with proteins, some aldehydes react chemically with amino groups to form Schiff bases. The reaction of aldehydes or ketones with cysteine to form a thiazolidine is reversible with heating, particularly at low pH.

Interactions with Carbohydrates

Binding of flavors to starch has been the subject of extensive study. (*7*) Different starches show varying flavor binding capacity. For example, starches with a low amylose content (e.g., tapioca) have a weak binding capacity, while those with a high amylose content (e.g., potato or corn) have a greater one. Starch is capable of forming inclusion complexes due to its helical structure, whereby hydrophobic regions exist in the inside of the polymer in which lipophylic flavors can be retained. Alcohols, aldehydes, acids, esters, terpenes, pyrazines, and other classes of flavors have been investigated (*5,6*).

Simple sugars often serve as carrier substances for flavors. In aqueous systems, there are conflicting results regarding the degree of interactions, which are presumably physical. In dry systems, a number of flavoring substances were shown to bind weakly to glucose, saccharose, and lactose (*6*).

With polysaccharides such as guar gum, alginates, agar, cellulose, and methyl cellulose, volatile flavoring substances (e.g., acetaldehyde, ethanol, diacetyl, ethyl acetate, and 2-hexanone) were shown to bind at varying degrees of strength. In general, increased concentrations of these polysaccharides causes a decrease in aroma and taste intensity. Since polysaccharides are often utilized as gelling or thickening agents in processed foods, the type and amount used may have a significant impact on balanced flavor perception. (*5,6*)

While beyond the scope of this book, encapsulants such as cyclodextrins easily form complexes with lipophilic volatiles in water. Similar to starch, cyclodextrins form inclusion complexes, in which the inner hydrophobic core of the glucopyranose ring retains lipophilic flavors. Reported uses of β-cyclodextrins are encapsulation of flavors, removal of off-flavors, and debittering of citrus juices.

Interactions with Lipids

Interactions of flavors with lipids are usually related to partitioning phenomena, or the relative amount of flavor solubilized in the lipid and water phases. The predominant lipids in foods are fats and oils, which consist predominantly of di- and triglycerides. Triglycerides can "bind" or solubilize considerable quantities of lipophilic and partly-lipophilic flavoring substances. The binding capacity of solid fats is lower than oils. The quantity of bound flavoring substance is also dependent on the fatty acid chain length and degree of unsaturation in the triglyceride composition (*8*). For instance, triglycerides with long-chain fatty acids bind less ethanol and ethyl acetate than those with short-chain fatty acids. Triolein—a triglyceride that contains only unsaturated oleic acid—binds more flavor than tripalmitin and trilaurin, both which contain only saturated fatty acids.

In lipid-water mixtures and emulsions, flavoring substances are distributed between the lipid and water phases as a function of their structure, the temperature, and the type of lipid. The amount of flavor bound by fat or oil is dependent on the chain length of the volatile compound within a homologous series. This implies that the concentration in the gas phase reduces as the chain length increases (*8*).

In food systems with lower fat levels, altered flavor-ingredient interactions produce a different flavor release behavior. Considerable interest in this phenomena has resulted during the development of low-fat and fat-free foods. Many volatile flavoring substances exhibit a lower vapor pressure in lipids, and therefore a higher odor threshold, than they do in aqueous systems. Reducing the amount of fat has the effect of raising the equilibrium vapor pressure of the flavorant, and changing its time-intensity release profile. Consequently, volatile flavors cannot be retained in the food matrix and are released immediately, resulting in a strong but quickly dissipating flavor impression.

Complex System Interactions

Thus far, flavor interactions have been described for individual classes of food components. However, most foods are mixtures of several ingredient classes. Initial attempts have been made to explore interactions with two or more components. For example, the amount of hexanal and 2-hexanone bound to soy protein plus guar gum, or guar gum plus an emulsifier, corresponds to that bound by the individual components. If all three components are mixed, significantly more binding of flavor is observed (*8*). Preliminary evidence has been demonstrated for the potential use of a protein-stabilized W/O/W emulsions for the retardation of flavor release (*9*). The combination of protein or starch with monoglyceride emulsifiers was shown to bind less flavor than protein or starch in the absence of emulsifier (*8*). Clearly, more studies need to be done in this area to understand the total interactive system.

Conclusions

Previous studies of flavor–food interactions have mainly focused on simple model systems using individual ingredients. The next step in this field is to systematically piece together collections of ingredients to explore the perspective of the food system as a whole. The systematic development of food products with acceptable flavor quality will be possible only when the flavor-binding behavior of food ingredients has been thoroughly elucidated. As described in the first chapter, future modeling studies also need to be expanded to include a direct link between the chemical data and the perceptual response of the sensory organism to provide a more complete understanding of flavor interactions.

Literature Cited

1. Maier, H. G. *Angew. Chem.* **1970**, *9*, 917.
2. Solms, J.; Osman-Ismail, F.; Beyeler, M. 1973, *Can. Inst. Food Sci. Technol. J.* **1973**, *6*, A10.
3. Le Thanh, M., Thibeaudeau, P.; Thibaut, M. A.; Voilley, A. *Food Chemistry* **1992**, *43*, 129.
4. Overbosch, P.; Afterof, W. G. M.; Haring, P. G. M. *Food Rev. Int.*, **1991** *7*, 137.
5. Plug, H.; Haring, P. *Food Quality and Preference* **1994**, *5*, 95.
6. Matheis, G. *DRAGOCO Report* **1993**, *38* (3) 98.
7. Solms, J. In *Interactions of Food Components*, Birch, G. G.; Lindley, M. G., Eds.; Elsevier Applied Science: London, 1986, pp. 189-210.
8. Matheis, G. *DRAGOCO Report* **1993**, *38* (4) 148.
9. Dickinson, E.; Evison, J.; Gramshaw, J. W.; Schwope, D. *Food Hydrocolloids*, **1994**, *8*, 63.

PERSPECTIVES

Chapter 1

Perspectives on the Effects of Interactions on Flavor Perception: An Overview

Derek G. Land

Taint Analysis and Sensory Quality Services, Loddon, Norwich NR14 6JT, United Kingdom

Flavor is essentially a perceived attribute which results from interactions between a living organism and certain chemicals released from food or beverages. It is not only a physico-chemical property of the molecules, the matrix and the amounts released, but also of the biology of the receiving, responding organism. The key concepts of flavor perception and response, some slightly speculative, will be outlined as an overview to flavor interactions in food systems.

The subject of flavor release and binding is a topic of recent interest, particularly in relation to new product development efforts which use novel ingredient systems. This chapter draws attention to the aspects of flavor interactions which are either emerging issues, or deserve further research. Alternatively, it does not attempt to provide a comprehensive review of the factors which influence the manner in which natural or added flavor substances interact with bulk foods. For this, the reader is referred to recent reviews of the subject by Solms & Guggenheim (*1*) and Matheis (*2,3*).

Biological Aspects of Flavor Perception

Flavor is the combination of sensations from taste stimuli dissolved in saliva, and retro-nasal odour stimuli in air delivered backwards into the nose from the mouth on chewing, but mainly on swallowing. These stimuli (chemical flavors or flavor substances) are released from food or drink in the mouth. The sensation is *not* the same as that arising from odor stimuli smelled only *ortho*-nasally by sniffing, although the smell sensation (and others, e.g., appearance) will influence the subsequent flavor perception; such effects result from biological cognitive interactions.

0097–6156/96/0633–0002$15.00/0

Taste and Retro-Nasal Smell. When food is put into the mouth, chemicals from the food surfaces dissolve in the saliva and diffuse through a mucous layer into contact with several types of sensory receptors on the tongue and in the mouth. Those chemicals which are sufficiently volatile to pass into the air in the mouth do so and are positively pumped retro-nasally into the nose perhaps by chewing but mainly by the act of swallowing. Evidence for this is from preliminary, unpublished tests (Land) with a soap-film flowmeter that have shown that on swallowing food, a small volume (5-15mL) of air is expelled from the nose at the time when the epiglottis has momentarily closed the trachea between breaths to prevent ingress of the food or drink during swallowing. This pulse of air can only be that which was in close, masticatory contact with the food or drink in the mouth immediately before swallowing, and will contain the volatile substances which produce the retro-nasal odor element of flavor. The odor-containing air is therefore not just being continuously diluted in a large flow of expired air during mastication, but is accumulating in a restricted pocket for undiluted expulsion into the nose on swallowing while there is no expiratory air flow. The swallow is always immediately followed by expiration, which will push the pulse upwards towards the olfactory cleft and out through the nares, perhaps with the initial velocity of the pulse producing selective diversion into the olfactory clefts before the major expired volume takes the lower route of least resistance well below the clefts. As normal respiration continues during mastication, the mouth must be closed off from the trachea to prevent food or drink entering the lungs during inspiration, although there may be some release of air from the mouth into expired air during chewing. Thus odorous volatile substances will then redissolve in the mucous layer of the olfactory epithelium, interact with and stimulate receptors.

Time Effects. Both processes occupy a finite time between initial mouth-approach to removal of direct stimulus source by swallowing; the time-base will vary with physical form and temperature of the food or drink and with the eating (e.g. meat could be chewed for 30 s) or drinking (2 s in mouth) behavior of the individual consumer. The time sequence of this model would reduce differences caused by different volatilities on rate of release or degree of equilibration. The dynamics of this pulse have never been explored, but should be. The model also has major consequences for interpretation of expired air and mouth flushing measurements of flavor release (*4, 5*).

Central Cognitive Interactions. All the receptor cells produce a range of time-varying and very complex neural signals which will vary with cell location and activity as well as with molecular properties of both stimuli and matrix, stimulus concentration and rate of concentration change. This battery of signals are then integrated by the brain into a wealth of previously experienced patterns of stimuli and current contextual expectations. The resultant flavor percept cannot be measured directly, but can only be expressed as behavior, which we do measure. It is inappropriate here to explore further these little-understood pathways, but it is very important to recognize that any human behavioral response to a flavor percept is usually very far removed from being a simple, one-to-one-stimulus response relationship, ever under the relatively controlled conditions of sensory analysis with a trained panel. There is enormous scope for a response to the same stimulus to be modified in different ways and contexts, and at different times in the same individual, and individual people can differ greatly at physiological and experiential levels. This is summarized for odor by: "Substances do not smell, they are smelt" (*6*), which means that what is smelled is not only a function of the stimulus, but of the smeller and the context. It is one reason why responses to the same flavor substance can differ greatly within and between individual people. I shall not dwell further on this except to comment that, while experimental flavor stimuli can be presented in a simplified context (e.g. tasting blind, or as simple solutions), real-world perception of flavor is *always* within a context that creates its own expectations. Flavor chemists who forget this are in peril of drawing false conclusions.

Food System Factors

I shall now focus on some of the physico-chemical, and to some extent within-food, biological interactions which influence the process of release of stimulus molecules from food into the mouth, i.e., during mixing with saliva and mastication. Almost without exception, foods and many beverages are complex mixtures of major and minor components, including indigenous and added flavor substances; most consist of at least two phases and often contain some traces of lipid material. To my awareness, relevant data on release of taste and odor substances is from simple model system studies which do cast some light on what happens to food in the mouth; however, few approach the complexity of real food, and some of what follows is speculation.

Composition of Matrices. The composition of the food matrix in which flavor substances are present, or in some cases in which they are biologically formed when the food is disrupted by processing or mastication, undoubtedly can play a very significant role in what is perceived as flavor. Recently, this subject was thoroughly reviewed by Solms & Guggenbuehl (*1*), and covers many known aspects of "binding" of flavors by e.g., starches and other carbohydrates, proteins, gums and thickeners, lipids and even purines. All of these food ingredient components reduce the proportion or rate of release of "free" flavor substances by some type of physical interaction, which range from bound ligands or clathrates between specific molecules to a simple viscosity increase which greatly limits flavor diffusion rates in the matrix. Some effects of apparently minor substitution can be surprisingly large. King (*7*) used sensory analysis to show very marked differences between five vanilla flavors in four different ice cream bases; the effect of base variation within each flavor was very striking, and emphasizes what flavorists have long known – many different flavor formulations are required to give similar perceived effects in even a narrow range of base matrix formulation. In this study, the butterfat level was constant at 12.5%, but even small variations in amount or type of fat would have produced large variations in perceived flavor using any one of the vanilla flavors. The familiar flavor problems from substitution of fat with either protein/carbohydrate-based mimetics, or non-digestible "fat substitutes" result directly from the different physico-chemical properties of the matrix. The problems of consumer acceptance are likely more than physico-chemical, due to cognitive interactions from sensation, expectation and socio-economic influences. The chemist's role in such new product development is only one element of a multi-dimensional evolving goal.

Food Factors which Influence Release of Flavor Substances. Volatile water-soluble substances which contribute to flavor can only do so when they are released from food into the air and aqueous saliva phases in the mouth; those in air also have to be reabsorbed into the aqueous mucous layer in contact with the sensory receptors within the nasal cavity. The physical laws which describe the processes of diffusion and the equilibration concentration ratios are understood for simple single-phase bulk systems such as water or oil (e.g., *8*), although there is much less data for many of the solid materials which are present in foods, and almost none for saliva and mucous. The processes and kinetics involved have been comprehensively reviewed (*9-11*).

However, although there are simple food ingredients, e.g., sucrose, it is extremely rare for any food to be a single substance or even for beverages to be a single aqueous phase. The vast majority of foods, whether "natural", processed or formulated, consist of three or sometimes four phases: solids, which can be amorphous, crystalline or glasses; lipophyllic or hydrophilic liquids; and gas dispersions (e.g., as in a mousse). Although beverages are mobile liquids (many with suspended solids) almost all foods are semi-solids as consumed, with only part of any liquid phase as free liquid in which available diffusion data can be used. These phases are rarely present as bulk component masses, but are stabilized by being very finely inter-distributed, e.g., as the cellular structure of plants or animals, or as emulsions, foams or membranes. This provides a further phase dimension, for it is now well-established that interface properties can be quite different from those of the respective bulk phases, and the presence of trace concentrations of solutes has a marked effect on those properties, e.g., detergents in water, trace elements in silicon chips. I shall illustrate this with some earlier data on simple equilibrium model systems.

Equilibrated Systems. The simplest models, used for decades now, are those in which a liquid, e.g., water or oil, containing a flavor-substance at a perceptually relevant low concentration is allowed to equilibrate with air to simulate release from food, and the concentration in the air (headspace) is then measured. Most published data at realistically low concentrations show that such systems do obey Henry's Law, which states that at equilibrium, the air concentration is proportional to the concentration in the liquid phase. For example, allyl isothiocyanate (AITC), a major flavor component of mustard, was equilibrated for 15-20 min. at 20 °C in five different media over a wide concentration range. Differences in vapor pressure for AITC can then be observed for the various media as shown in Figure 1 (Land, D. G.; Hobson-Frohock, A.; Reynolds, J., unpublished data). Deviation from linearity for Henry's Law occurs only at very high concentration where solubility of AITC in

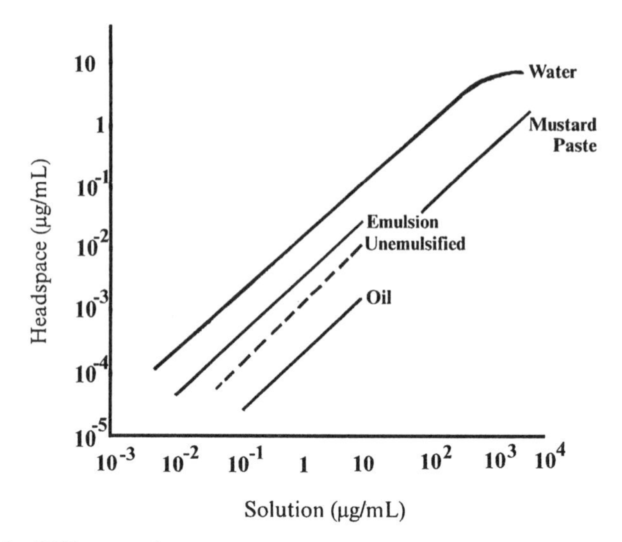

Figure 1. Differences in vapor pressure of allyl isothiocyanate (AITC) above various media in relation to solution concentration.

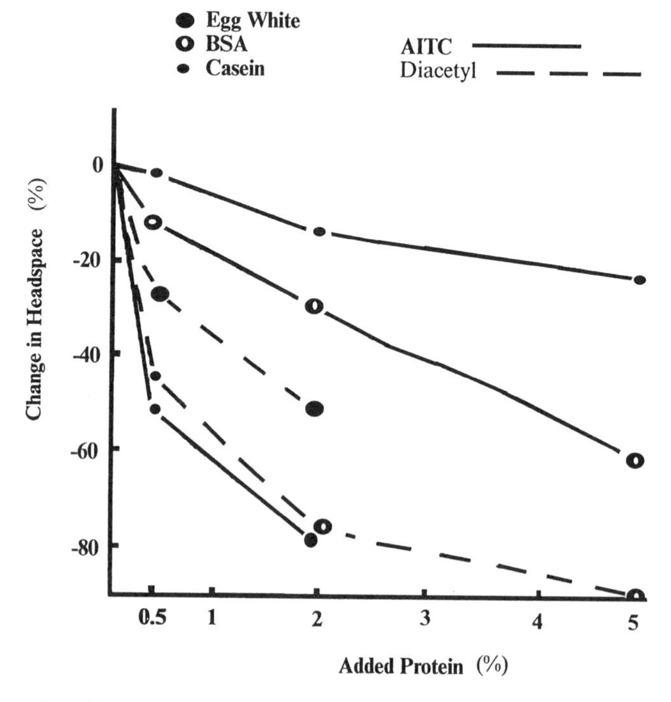

Figure 2. Effect of native proteins on vapor pressure of allyl isothiocyanate (AITC) and diacetyl in water.

water becomes limiting. The plots of the various AITC solutions have almost identical slopes, but quite different partition coefficients. This implies that different liquid phase concentrations must be produced to obtain the same concentrations of AITC released in air, at the same perceived aroma intensities. The concentration required in groundnut oil is 100-times that in water, and reflects the oil solubility of AITC. Mustard paste (a food condiment) requires 14-times more AITC than water, and this "excess" must be held by the 33% of non-aqueous components in the paste, which consist of approximately 8% protein, 8% oil and 10% carbohydrate, the latter composed of mostly simple sugars, with some starch and fiber. This composition is not dissimilar to that of many foods as consumed. One conclusion to be drawn is that the suspended solids and oil in the paste "bind" the excess AITC.

In Figure 1, two of the liquid phases consist of water, oil and emulsifier of identical composition, but one variable was allowed to remain as two phases with a fluffy interface where the emulsifier accumulated, and the other was shaken to form an emulsion. In the emulsion situation, more AITC is released into the headspace than in the two-layer state. A possible explanation is that AITC is more soluble in, or adsorbed on, the surface of solid emulsifier than when it is widely spread as a monomolecular layer at the very much larger interfacial surface of the emulsion; similar results were obtained whether the AITC was added in the oil or in the water phase. In this case mastication of the unemulsified preparation could have produced some emulsification, and thus increased the amount of flavor released and the consequent perceived intensity, although this was not tested by sensory means.

Binding with Proteins. Binding of volatile substances by proteins has long been established (e.g., by casein and whey protein (*12*)) although many of the earlier studies were carried out at unrealistically high concentrations of volatile substances, mainly for ease of analysis. An example using diacetyl (*13*) at sensorily relevant levels (0.5 μg/mL aqueous solution) shows the extent of flavor binding by proteins even at very low (0.5%) protein concentrations (Figure 2; Land, D. G.; Reynolds, J., unpublished data). Headspace was measured by gas chromatography and detected by electron capture. The effect is very sensitive to protein type and conformation (*14*). However, bulk composition and chemical nature of the ingredients of food is not the only factor which influences release of flavors.

An analogous but opposite effect which did not obey Henry's Law was obtained (*15*) by using dimethyl sulfide at sensorily relevant levels as the solute (Figure 3). The non-linearity presumably reflects increasing saturation of a small number of "binding sites" on the emulsion interface. A similar but smaller effect was found with diacetyl (Land, D. G.; Reynolds, J., unpublished data) (Figure 4). It should be noted that these observations have recently been challenged (*10*) and independent repetition is required. However, if confirmed, these effects could be very significant at the very low levels at which most important flavor-impact substances occur in the complex lipid membrane-like interfaces which form a major portion of most foods.

Type of Lipid. Unfortunately, there is little published quantitative data at sensorily realistic flavor-substance levels on the effects of different lipids or lipid substitutes

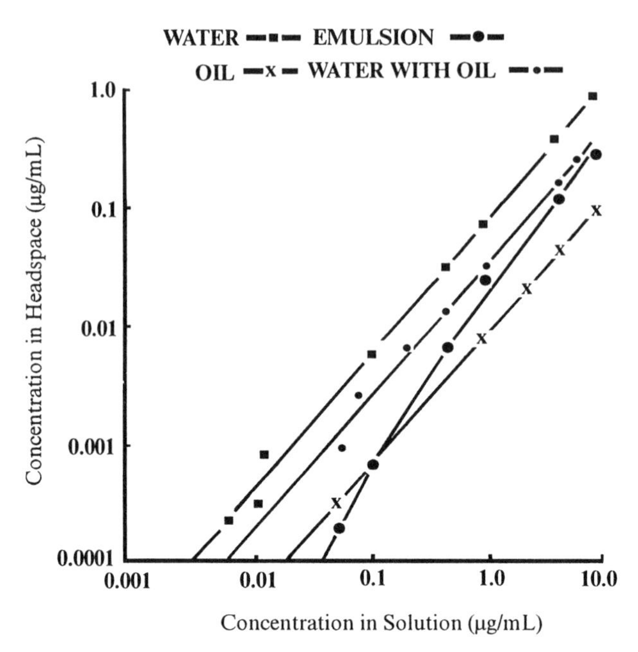

Figure 3. Differences in concentration of dimethyl sulfide in the headspace above various media. (Adapted from ref. 15. Copyright 1979 Applied Science Publishers).

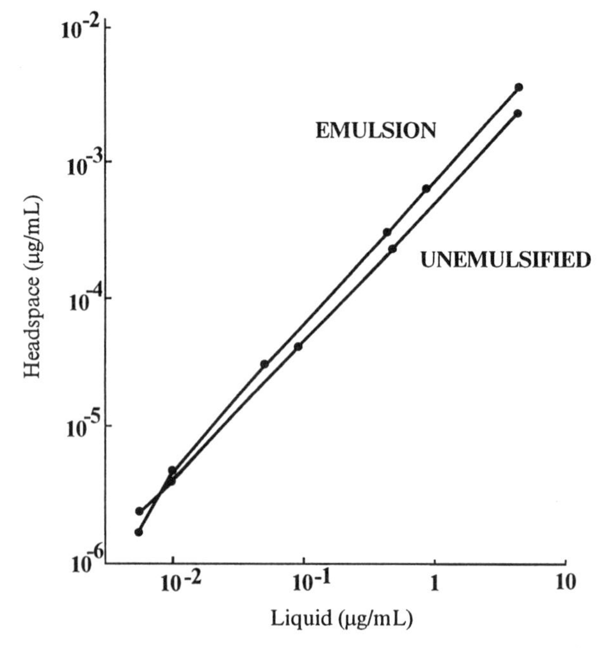

Figure 4. Vapor concentrations of diacetyl over water/groundnut oil.

on flavor release, although much work has been done but not published. There are major effects due to change of state on melting in bulk phases, e.g. in chocolate. Such changes may be caused by strongly held inclusions (clathrates?) within solid crystal structures, with release only on melting or dissolution of the crystal structure, or they may release from the more rapid diffusion of small molecules through liquids vs. through solids. Type and configuration of lipid will thus have effects on release. However, lipid type, structure and location is likely to have far more effect by influencing the surface properties of interfacial films and membranes.

Dynamics of Flavor Release. As food is rarely chewed and held in the mouth for more than 15-20 seconds, and beverages are rarely held for more than 5 seconds, it is clear that full equilibration with air or saliva is very unlikely to be achieved before swallowing. This means that the rate of release will determine the extent to which equilibrium concentrations of flavor substances in the air in the mouth are actually achieved. Actual in vivo measurements for high concentrations of less-potent flavors were not practically possible until a few years ago, and are still not possible for the interesting and important potent flavors which occur at very low concentrations. The topic has been extensively reviewed by McNulty (*9*) and by Overbosch et al (*10*). The main conclusions are that rate of release is influenced by both the flavor and the composition of the medium, e.g., that 2-hexanone is released twice as fast from water as from oil at 37 °C, and the rates appeared to be slow, although times to saturate the air were not given. Slow equilibration of air does not always apply, because diacetyl was found to reach 90% saturation in air at 15 seconds, 37 °C for sensorily significant concentrations in water (Land, D. G.; Reynolds, J., unpublished data). Lee (*16*) showed 90% of maximum diacetyl in air concentrations, reached in 22 and 28 seconds from palm stearin and olein, respectively, at much higher concentration. As it is rare to find a food flavor which results from only a single flavor impact substance, one could conclude that the intensity and perceived quality of most food flavors should change markedly with the time that food remains in the mouth. However, increasing amounts of time-intensity sensory data show that, excluding bitterness and aftertastes, maximum intensity and full flavor quality typically develop within 10 seconds of food being chewed and swallowed. In practice, differential release of flavor does not appear to occur to any great extent in normal food consumption; one possible explanation is proposed in the mechanism of retro-nasal release above. Differential release often is used by trained and experienced flavorists, but they employ special techniques to enhance such effects. If the supposition is correct that differential release is practically irrelevant, then some of the concepts upon which current models are based are not correct. I suggest that such errors arise from two main sources – the use of bulk-phase media as models, and of expiratory flow as the major means to transfer volatile flavor in air from within the mouth to the nose.

Bulk-phase Models. In any bulk phase, release of solute into surrounding phase, whether liquid (e.g., saliva) or gas (e.g., air) is rate-limited by diffusion to the boundary driven by the concentration gradient. As already stated, real foods rarely have bulk fat/oil or solid phases for they are usually finely dispersed either as cellular structure, e.g., meat or potato, or by processing, e.g., dairy products. Even in high-fat (up to 40%) foods such as avocado, the lipid is finely dispersed and will have a very large interfacial area; the bulk-phase model will therefore not apply. Generally, in food as eaten there will be little or no bulk lipid, for it is present as mono- or bi-layers in membranes, or as fine globules in cells or emulsions. This implies that disruption/exposure of any new surface will be followed by very rapid "saturation" of the relatively small volume of fresh saliva and air bought into contact, because the unit solute "cells" or particles are very small and diffusion distances will be on the order of micrometers. To the author's knowledge, this model of lipid, protein and other food, saliva and mucous as thin flavor retaining and releasing membranes, or functionally interdependent interface layers, has not been explored or adequately modeled. However, it is compatible with current theories of lactoglobulin-like proteins as a cross-membrane stimulus transfer mechanism in olfaction (*17*). This type of model is not far removed from the lipid film puncture theory of odor receptor simulation of Davies (*18*); the latter was based on Langmuir trough data regarding the effect of trace odorants on the surface properties of monomolecular lipid films on water, and is probably worth re-examination considering recent knowledge.

Conclusions

Many food components, whether major base constituents, minor but highly functional components, or flavor substances themselves, can play substantial parts in the interactions which determine the flavor of food or beverages. Of these, lipids undoubtedly play a key role in food flavor, partly as a source of flavor substances and partly as a sink or reservoir but mainly through their very extensive role as membranes and interfaces. However, it is essential to remember that these physico-chemical phenomena of the food itself are not the sole determinant of flavor response behavior. Interactions with the person will play an important role, whether it is at the eating behavior level and mechanism of flavor in the mouth, or at higher cognitive levels of interaction with experience, expectations or attitude. This chapter briefly outlines some of the micro-environmental factors in the composition and structure of foods which might have an important influence on the way in which flavor substances are held in food, released in the mouth and contribute to overall flavor by retro-nasal odor perception. It provides some possible explanations for flavors being perceived as different in various media, and suggests two somewhat different perspectives for future modeling and investigation of food flavors.

Literature Cited

1. Solms, J; Guggenbuehl, B. In *Flavour Sci. Technol.*, Weurman Symp. 6th; Bessiere, Y.; Thomas, A. F., Eds.; Wiley: Chichester, UK, 1990, pp. 319-335.
2. Matheis, G. *DRAGOCO Report* **1993**, *38* (3) 98-114.
3. Matheis, G. *DRAGOCO Report* **1993**, *38* (4) 148-161.
4. Linforth, R. S. T.; Taylor, A. J. *Food Chemistry* **1993**, *48*, 115-120.

5. Delahunty, C. M.; Piggott, J. R.; Connor, J. M.; Patterson, A. In *Trends in Flavour Research*, Maaarse, H.; van der Heij, D. G, Eds.; Elsevier: Amsterdam, 1994, pp. 47-52.
6. Harper, R.; Bate Smith, E. C.; Land, D. G. *Odour Description and Odour Classification*; Churchill: Edinburgh, 1968.
7. King, B. M. *Lebensm. Wiss. Technol.* **1994**, *27*, 450.
8. Buttery, R. G.; Ling, L. C.; Guadagni, D. G. *J. Agric. Food Chem.* **1969**, *17*, 385.
9. McNulty, P.B. In *Food Structure and Behaviour*, Blanshard, J. M. V.; Lillford, P., Eds.; Academic: London, 1987, p. 245.
10. Overbosch, P.; Afterof, W. G. M.; Haring, P. G. M. *Food Revs. Int.* **1991**, *7*, 137.
11. Plug, H.; Haring, P. *Trends Food Sci. Technol.* **1993**, *4*, 150-152.
12. Reineccius, G. A.; Coulter, S. T. *J. Dairy Sci.* **1969**, *52*, 1219.
13. Land, D. G.; Reynolds, J. *Flavour '81*, Schrier, P., Ed.; DeGruyter: Berlin, 1981, p. 701.
14. Dumont, J. P.; Land, D. G. *J. Agric. Food Chem.* **1986**, *34*, 1041.
15. Land, D. G. In *Progress in Flavour Research*, Land D. G.; Nursten, H. E., Eds.; Applied Science: London, 1979, p. 53.
16. Lee, W. E. *J. Food Sci.* **1986**, *51*, 249.
17. Godovac-Zimmerman, J. *TIBS* **1988**, *13*, (2), 64.
18. Davies, J. T. *J. Theoretical Biology*, **1965**, *8*, 1-7.

Specific Food Component Interactions

Chapter 2

Implications of Fat on Flavor

L. C. Hatchwell

NutraSweet Kelco Company, 601 East Kensington Road, Mt. Prospect, IL 60056–1300

Fat plays an important role in the flavor perception of foods. It influences temporal profile, flavor impact, perception of flavor notes, and order of their occurrence. Fat replacers are composed of proteins and carbohydrates, which interact differently with aroma chemicals than fat does. Therefore, flavor challenges are faced whenever fat is reduced in a food product. The role of fat in flavor perception is reviewed, and the interaction of fat and fat replacers with aroma chemicals is discussed. The resulting effect on applications and some solutions to obtaining the appropriate flavor profiles are proposed.

One of the most challenging aspects of reduced-fat foods is the development of good flavor. Traditionally, the impact of fat on flavor perception was not well understood by the product developer. Understanding the functionality of fat as it pertains to flavor delivery and character facilitates the development of reduced-fat foods with optimum sensory qualities. This chapter will discuss fat and flavor interactions and some solutions to obtaining the desired flavor profiles.

Functional Aspects

Fats influence all aspects of food perception, including appearance, texture, mouthfeel and flavor. For appearance: sheen, opacity, oiliness, crystallinity, color development and color stability are important. Texture attributes include viscosity, tenderness, elasticity, cuttability, flakiness, emulsification, and ice crystallization. Mouthfeel can encompass cooling, lubricity, thickness, meltability, cohesiveness, mouth coating, and adhesiveness, all of which may contribute to the complex known as creaminess. Fat influences flavor attributes such as aroma, flavor character (fatty/oily, dairy), flavor masking, flavor release, and flavor development. In addition, fat plays an important role in the processing and preparation of foods, and

0097–6156/96/0633–0014$15.00/0

in storage stability. When formulating any full-fat or reduced-fat food products, developers must be cognizant of the various functions of fat (*1*).

Effect of Fat on Flavor Perception. Flavor compounds are inherent in lipid ingredients, whether they be desirable flavors such as those of milkfat, lard, olive oil, etc., or undesirable, such as those of emulsifiers. Flavors with "fatty" sensory attributes come from a variety of different aroma chemicals (fatty acids, fatty acid esters, lactones, carbonyl compounds and many others). These aroma compounds may contribute to give the sweet, buttery, creamy, "rich" flavor that comprise the entire range of dairy products, or combined differently, could give the notes peculiar to lard. Fats act as precursors to flavor development by interacting with proteins and other ingredients when heated. Examples of this are the flavors that develop during baking and roasting. A clear example is the flavor that comes from the combination of milk fat heated with sugar and eggs which results in the rich, caramellic notes of a full fat vanilla ice cream. In addition, fat participates in fermentation to result in the desired flavor components of cultured dairy products, particularly cheese (*2*).

The ability of fat to mask off-flavors may be due to solubility. Off-flavors may not normally be perceived in full-fat systems because most are fat-soluble and are at or below threshold levels. However, in the absence of fat, the vapor pressure of the aroma chemicals responsible for these off-flavors may be increased. This results in a very intense perception of the off-flavor.

Fat provides mouthfeel and richness. It interacts with flavor components to provide a specific sensory balance. In most food products, flavor components partition into the aqueous and lipid phases of the food, resulting in a balanced flavor profile (*3*).

Flavor release is a critical factor governing smell and taste. The majority of aroma chemicals are at least partially soluble in fat (*4*). This means that they are dissolved to some extent in the lipid phases of food, releasing the flavor slowly in the mouth and resulting in a pleasant aftertaste. Altering the type of fat or total fat content of foods affects the rate and concentration at which food flavor molecules are volatilized during consumption (*5*).

Removing any significant amount of fat (around 25% or more) from a product changes the flavor profile. As the concentration of fat is further reduced, the flavor challenges are increased significantly. In all cases of fat reduction, some flavor balancing is needed. As more fat is removed, the differences become much more apparent and the challenges to the product developer increase.

When fat is removed from a formulation, the only ingredients available to replace it are water, protein, carbohydrates, minerals or air. Even if nothing new is added to the formula, these items automatically increase proportionally. Each of these components interacts differently with flavor than fat does. A combination of these ingredients may mimic part of fat's function, but they cannot totally replace its functionality.

It is important to remember that fat and water are solvents for aroma chemicals (*6*). Proteins and carbohydrates absorb, complex, and bind with aroma chemicals; they never act as solvents. A fat-soluble flavoring (for example, lemon oil) can be

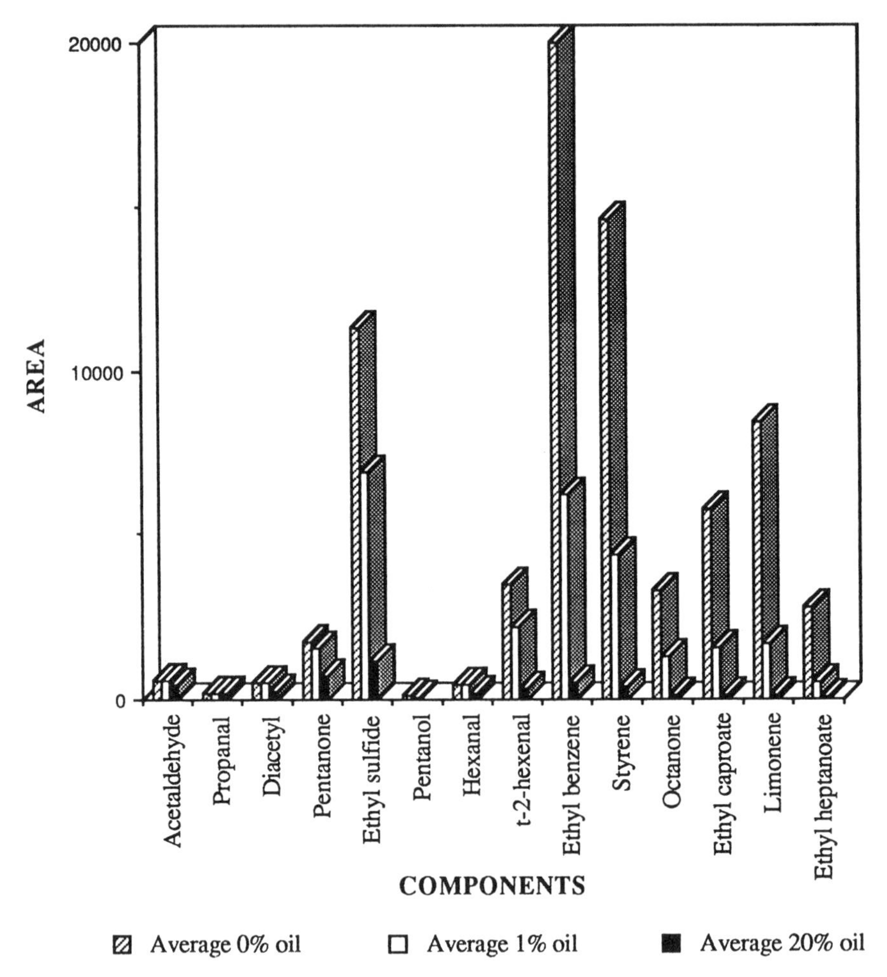

Figure 1. Interaction of selected aroma chemicals with 0%, 1%, and 20% fat. (Reproduced with permission from ref. 8. Copyright 1994 Institute of Food Technologists.)

solubilized in one of two ways — in oil or as an emulsion. Imagine lemon oil in a closed container in a water emulsion, and another container with oil and lemon oil. Since the lemon oil is hydrophobic, its molecules will not be held in solution as much by water molecules as they would be by those of a surfactant or oil. Therefore the lemon oil is more likely to surface and volatilize in a water system than in an oil system. In the closed container, the headspace of the volatiles will build up. When the container is opened and smelled, the water/gum blend will appear to be stronger in odor than the oil blend. It will be perceived as sharper and harsher. Most volatile components have a tendency to be more oil-soluble than water-soluble. Reduced-fat systems inherently have less fat and more water. Therefore, aroma chemicals may be perceived as strong and unbalanced.

A small amount of fat can be utilized effectively to further flavor perception. As little as 1–2% fat is enough to affect the flavor components (7) and make a big difference in flavor perception. Figure 1 shows the headspace concentration of aroma chemicals in the presence of fat (*8*). The more water-soluble chemicals, such as acetaldehyde, propanal, diacetyl, pentanol, hexenal, do not show much change in interaction when fat is added. The fat-soluble chemicals, ethyl sulfide, ethyl benzene, styrene, and limonene exhibit dramatic differences in the amount of volatiles in the headspace when only 1% fat is added. Pragmatically, this means that a small amount of fat can be manipulated to alter flavor perception and result in a profile that is more similar to full-fat foods.

Not only does fat affect the intensity of flavor perception, it influences the temporal profile (*9*). Temporal profile is the timing of the perception of aroma once food has been placed in the mouth. In a full-fat food, the initial impact and intensity of the flavor is suppressed. The flavor intensity then increases and plateaus into a balanced flavor profile which tails off into a pleasant aftertaste. The fat-free food, on the other hand, exhibits immediate impact, then severely diminishes. The immediate impact is not perceived as a balanced profile but manifests itself in a series of uneven and sharp flavor notes. Vanilla is a good example. In the full-fat ice cream, the bouquet reaches the nostrils on the way to the mouth. Once placed in the mouth, the fat melts, slowly releasing the flavor. In the fat free ice cream, the immediate impact of vanilla is characterized by a series of unbalanced and seemingly foreign notes: smoky, medicinal, alcoholic, beany, woody. This profile then fades, resulting in no aftertaste. This is perceived as unpleasant by the consumer.

Fat Replacers

A fat mimetic is a carbohydrate or protein that replaces one or more of the functions of fat (Table I). Carbohydrates often work by absorbing large amounts of water to increase perceived moistness. They can be a good source of dietary fiber. Some require special processing such as pre-making a paste or gel. Use of carbohydrate-based fat-replacers in products with very low fat levels does not yield optimal results. Examples of carbohydrate-based fat mimetics are a modified low-methoxy pectin, polydextrose, maltodextrins, and dextrins made from oat flour or potato starch.

Table I. Definitions and Examples of Various Fat Mimetics, Fat Replacers, and Fat Substitutes

Fat Mimetic *Mimics one or more functions of fat* Microparticulated protein Modified starch Cellulose and cellulose derivatives Gelatin Gums Dietary fiber
Fat Barrier *Retards absorption of fat during processing* Gums Cellulose and microcrystalline cellulose
Fat Extender/Sparer *Maximizes the effect of fat* Emulsifiers and emulsifier blends
Fat Substitute/Analog *Lipid-based ingredient with characeristics of fat, but with altered digestibility* Sucrose polyester and synthetic oils (0 Kcal/g) Structured lipids (caprocaprylobehenin, triacylglycerols) (5 Kcal/g)

Effective protein-based fat replacers are micro-particulated. The microparticulated proteins function by binding water, but to a lesser extent than carbohydrates. They provide hydrophobic sites that aid in emulsification of the remaining fats in the reduced-fat system. These functionalities result in improved mouthfeel, ice-crystal control, and foam stabilitzation in semi-solid food products, such as ice cream; increased tenderness and crumb quality in baked goods; and help retain moistness and retard rubberiness in reduced-fat cheeses. The functionality in cheese is due to the microparticulated protein's unique capability to be retained in the casein matrix of the curd.

An example of a fat extender or sparer is the entire range of emulsifiers. These are derived from fat but are used at low levels. Therefore, their calorie contribution is low. As fat is removed, emulsifiers are needed to aid in moisture absorption, emulsion stability, aeration or defoaming (depending on the system). Emulsifiers help maintain tenderness in a reduced-fat baked goods. Some can be used as release agents in the machining of reduced-fat crackers. Emulsifiers improve the eating quality and shelf life of reduced-fat foods.

A fat substitute is sometimes referred to as a fat analog (Table I). It replaces all of the functions of fat in a product with decreased or no caloric contribution. Fat substitutes are usually modified fats that are poorly absorbed. Because these substitutes are molecules whose physical and thermal properties resemble fat molecules, they can theoretically replace fat in all applications, even frying. An example of such an ingredient is sucrose polyester (Olestra™, Procter & Gamble), which has been recently approved by the FDA for snack food applications. Another is caprenin, a transesterified fat used to replace the fat in a chocolate coating.

All fat substitutes and fat mimetics interact differently with aroma chemicals than fat does (*10*). Figure 2 compares the relative vapor pressure (RVP) of a homologous series of aldehydes (*7*). The interaction of water and these aroma chemicals has a RVP of 1. Slendid™ (Hercules), Oatrim™ (Quaker Oats/Rhone Poulenc), Paselli™ (Avebe), and Stellar™ (Staley) are forms of carbohydrate-based fat replacers (pectin, oat dextrin, potato starch, corn starch, respectively). Simplesse™ 300 and Simplesse™ 100 (NutraSweet) are microparticulated proteins (egg albumin/casein combination, and whey, respectively). As the aldehydes increase in chain length, they become more oil soluble. This leads to less of the aroma chemical migrating to the headspace in the oil. It is clear that all of these ingredients interact with aroma chemicals differently than they do with fat. Carbohydrate-based fat replacers have little impact on the RVP, probably due to the fact that they exhibit no hydrophobic groups. Protein-based fat replacers, conversely, have hydrophobic sites and thus bind longer chain aldehydes. Even if a combination of these fat substitutes were used, the flavor interaction would be different than with fat (*7*).

Therefore, one can see how use of traditional flavors may result in aroma and flavor imbalance when fat replacers are substituted for fat. The reduction of fat causes traditional flavors to be stronger and have immediate impact. However, careful blending of fat replacers and modified flavors can bring the flavor closer to the full-fat system.

Remembering all the functions of fat on flavor perception—indigenous flavor, flavor precursors, flavor masking, flavor partitioning, and flavor release—what does the scientist need to take into consideration in the formulation of reduced-fat products?

Key Issues

Raw Material Quality. While changes in texture can be handled with a variety of fat substitutes and fat-replacement technologies, it is best to choose a fat-substitute that does not exacerbate the flavor difficulties. Raw material quality is one of the most important aspects that a scientist must take into consideration. Normally, windows of acceptability are established for ingredient quality. Reduced-fat products are less "tolerant" of off-flavors than full-fat products. Thus, there is a narrower range of acceptability. Any defects in raw materials become readily apparent. Off-flavors are more pronounced and the flavor of the raw materials used in the formulation are uncovered. For example, gums and starches develop stale, cardboard notes when exposed to light or aged too long; fruitiness and earthiness are inherent in sugar, depending on the source of manufacture; oxidized flavor and bitterness develop through the aging and storage of dry milk; and gelatin will

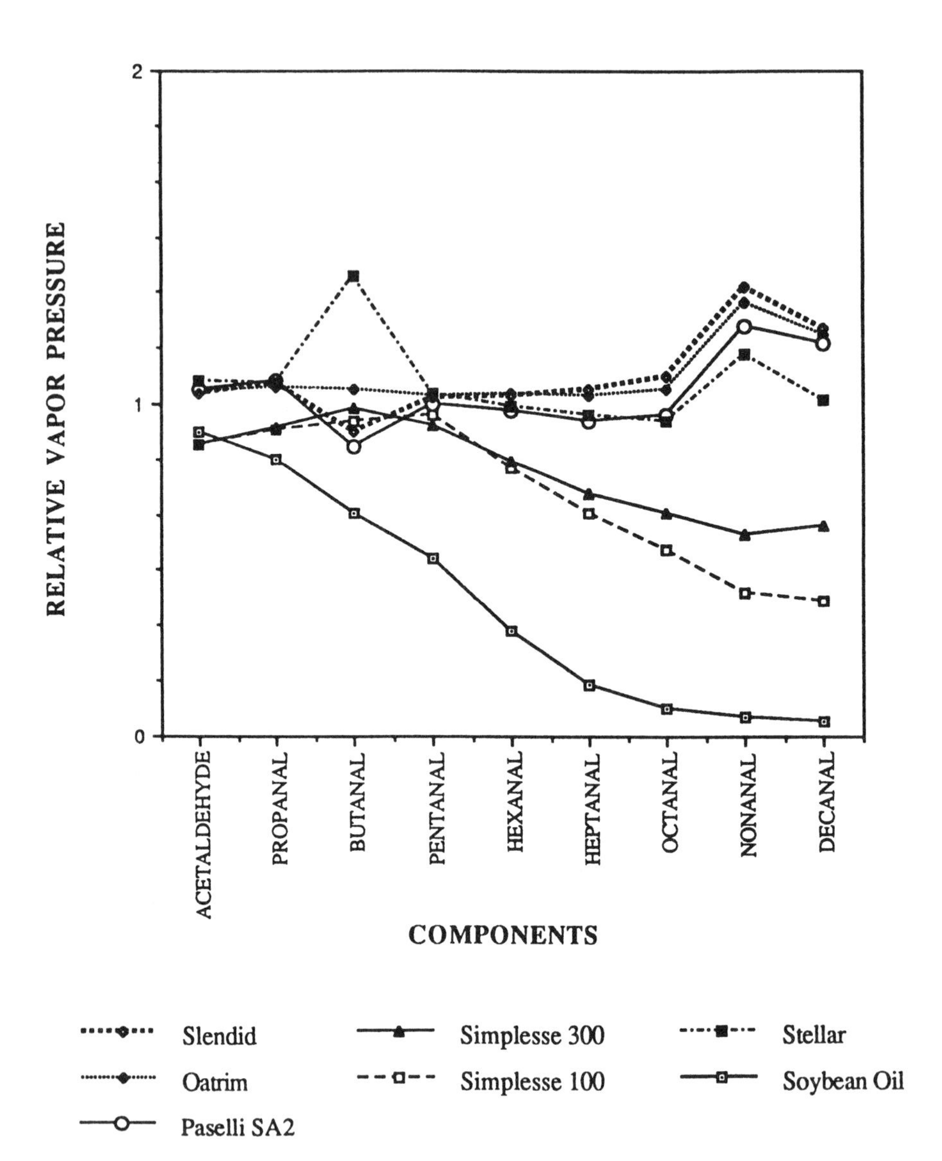

Figure 2. Interaction of fat mimetics with aldehydes. Commercial fat mimetics (ingredient composition): Slendid™ (low-methoxy pectin); Oatrim™ (modified oat flour); Paselli SA2™ (potato maltodextrin); Simplesse™ 300 (microparticulated egg albumin and milk); Simplesse™ 100 (microparticulated whey protein); Stellar™ (crystalline corn starch). (Reproduced with permission from ref. 7. Copyright 1994 Institute of Food Technologists.)

develop fishy notes when aged. In addition, high-amylose starches bind flavor into the amylose helix, resulting in diminished flavor perception.

Base Characteristics. Quality of the food system base is also very important. The pre-flavored base must be free of flavor defects. There is no such thing as the masking of off-flavors in a reduced-fat system. All defects will be perceived. In comparison, a base with no flavor is also undesirable. It is very difficult to provide an entire flavor profile via aroma chemicals without producing an imbalanced artificial tasting product. The base must have some of the desired flavor character on which to build. For instance, a dairy character should be built in via the ingredients in a reduced-fat ice cream mix.

The balance of sweetness and saltiness is critical. As water is increased in a product formula, perceived saltiness and sweetness decrease. When the sweetness and saltiness are adjusted, the increased salt and sugar effect the perceived balance of flavor. Bases must be complete before flavor development can begin. If possible, build clean bases with flavors that are inherent to the desired system. Then adjust the salt and sweetener concentrations.

Flavor Systems. Once a clean base with some of the desired character and appropriate balance of sweetness and saltiness has been developed, flavor systems can be investigated. Simply raising or reducing the usage level of a flavor does not create a balanced flavor. Skillful flavor chemists can modify the flavor formula to achieve a balanced flavor profile. It is advisable to obtain flavors from a variety of flavor companies. Each company has its unique technology for handling flavors for reduced-fat products. Combining these technologies can result in a complete and balanced flavor system.

As fat is decreased in products, especially liquid or semi-solid foods, the usage level of flavors needs to be lowered. Due to the increase in the concentration of water in the reduced-fat formulation, a large impact can be perceived from even low levels of flavor. As we saw before with the model systems, addition of small amounts of fat can greatly influence the perception of flavor quality and order of occurrence. Effective use of that fat is important.

Some select flavors enhance flavor perception. These include mouthfeel flavors, fat flavors, dairy flavors, caramellic flavors, and flavor modifiers/ potentiators. They provide flavor notes that normally develop when fat is present in the formulation.

It is possible to add some of these flavors into products prior to processing. They then function as precursors for the final flavor development. Examples are mouthfeel flavors added to ice cream; butter and caramellic precursors added to a caramel candy; or butter precursors added to a sauce or to a baked good.

Mouthfeel flavors are combinations of aroma chemicals that provide a feeling of fullness and flavor delivery. They help delay and prolong flavor impact. They are used as adjuncts and for main aromatic character. Fat flavors are mixtures of aroma chemicals that mimic the flavor of fat. Dairy flavors (sweet milk, cream, condensed milk flavors, and others) can help round off other flavors, when used judiciously.

All of these types of flavors are difficult to use properly. High concentrations are unpleasant. Consumers tend to either prefer or dislike these flavors. This is because the diacetyl and lactones used in the flavor formulations are perceived by a segment of the population as coconut, peachy, or green butter, even when used at minute levels. These flavors have a tendency to change their flavor profile during storage. Manufacturers need to be aware of this and to use fresh ingredients.

Alternatively, addition of appropriate amounts of enzyme-modified cream or butter prior to processing can provide precursors for flavor development, resulting in the desired caramellic notes needed in many reduced-fat products.

Flavor potentiators are added after a flavor blend has been balanced and accepted. In sweet products, such as ice cream and baked goods, maltol is an example of a flavor potentiator that helps balance and prolong flavor.

Conclusions

The effects of fat removal in most food products include flavor and aroma imbalance, changes in temporal profile, modification of texture and mouthfeel, awareness of off-flavors, changes in acceptability of raw material quality, shelf-stability and packaging interactions. There is no one solution to fat reduction, but a combination of various strategies will move the product under development closer to its target.

Use of reduced-fat foods in the consumer's diet will continue to be important as they strive for a healthful life-style. The development of great tasting reduced-fat products will continue to be a demanding challenge. Understanding the science behind flavor interaction in reduced-fat systems will help flavor chemists develop better flavors for these systems. Product developers must consider all aspects of the functionality of fat. Ingredients must be reconsidered and used in a different way than they have in the past. Since flavors must be used differently in reduced-fat systems, a variety of approaches and technologies must be tried.

There is an opportunity and a need for professionals in product development, food science and chemistry to work together to define the science of reduced-fat foods with full-fat attributes. This will result in reduced-fat products that still have the function and flavor of their full-fat counterparts.

Acknowledgments

The author wishes to thank Angela Miraglio for all her support with the editing of this text; Anne Hendrickson for conducting literature searches; Jo Milazzo and Ellen Tieberg for their assistance with layout; and all of the people at The NutraSweet Company who supported the development of Simplesse™ and reduced-fat products.

Literature Cited

1. Best, D. *Prepared Foods* **1991**, *160*, 72 - 77.
2. Ashurst, P. R. *Food Flavorings*; Blackie and Son, Ltd.: Bishopbriggs, Glasgow, Scotland, 1991.

3. deVor, H. *Food Industries of South Africa.* **1993**, (8), 48-49.
4. Forss, D. A. *J. Agric. Food Chem.* **1969,** *17,* 681-685.
5. Shamil, S., Kilcast, D. *Nutrition & Food Science* **1992,** *4,* 7-10.
6. Furia, T. E.; Bellanca, N. *Fenaroli's Handbook of Flavor Ingredients,* 2nd ed.; CRC Press, Inc.: Boca Raton, Fla, 1990, Vols. I and II.
7. Schirle-Keller, J. P., Reineccius, G. A., Hatchwell, L. C. *J. Food Sci.* **1994,** *59,* 813, 815, 875.
8. Hatchwell, L. C. *Food Technology* **1994,** *48,* 98.
9. Shamil, S., Wyeth, L., Kilcast, D. *Food Quality & Preference* **1991**, *3*, 51-60.
10. Schirle-Keller, J. P., Chang, H. H., Reineccius, G. A. *J. Food Sci.* **1992**, *57,* 1448-1451.

Chapter 3

Performance of Vanilla Flavor in Low-Fat Ice Cream

Ernst Graf[1] and Kris B. de Roos[2]

[1]Flavor Frontier, 731 Wright Street, Rathdrum, ID 83858
[2]Tastemaker, Nijverheidsweg 60, 3771 ME Barneveld, Holland

Fat removal from vanilla ice cream results in drastic flavor profile distortion and loss in vanillin intensity during storage. The chemical instability of low-fat ice cream results from both chemical reactions and physical interactions with proteins, starches and other hydrocolloids. Initial flavor perception, however, depends primarily on phase partitioning of the individual chemical components between water and oil. A nonequilibrium partition model was developed to accurately predict flavor performance in foods and beverages. The proposed physicochemical model describes the effect of fat level on the flavor profile and it calculates a reformulation factor for each chemical component of a compounded flavor to restore the original taste in a reduced fat product. It also allowed for the design of a novel cryogenic fat enrobement technology for a vanilla extract that cannot be reformulated. In this case we created a microenvironment for the flavor that mimics high-fat ice cream.

Consumer preoccupation with excess dietary fat has been steadily rising over the past decade (*1*). In 1993 the percent rating factor as greatest concern with fat content reached 54% and was exactly twice that for salt or cholesterol levels. A steady growth in low- or no-fat food products clearly reflects this consumer health awareness. Find/SVP pegs the US market for low-fat and/or low-cholesterol prepared foods at $15.7 billion in 1992, $23.5 billion in 1993, and forecasts $44.9 billion by 1997 (*2*). Of the 12893 new food products introduced in 1993, the low-fat segment scored a total of 577 or 4.4%.

Despite the widespread interest in total caloric reduction and fat removal, many consumers exhibit fairly sporadic purchasing behavior (*3*). The gap between actual eating patterns and the marked concern with healthy food arises primarily from the compromise in flavor of low-fat foods. In a recent survey of the relative

0097–6156/96/0633–0024$15.00/0

importance of five product attributes, shoppers ranked taste clearly above nutrition, price, product safety and storability. In the preface to *Food Technology's* 1994 top ten food trends (*4*) the author concluded that taste alone will reign as the most powerful criterion for food selection.

Demographic changes and increasing affluence in Western societies have resulted in a growing consumer demand for healthy food of outstanding quality. The significance of taste holds particularly true for the frozen dessert market. However, several sensory studies have demonstrated that the flavor of no-fat or calorie-reduced ice creams are unable to match their full fat analogs. The development of premium low-fat ice cream poses a serious technological challenge but also a competitive business opportunity to the sophisticated flavor chemist.

Calorie reduction in vanilla ice cream results not only in a rapid loss in vanillin intensity during storage, but also in drastic initial flavor profile distortion. The altered flavor performance in a fat-free or low-fat ice cream containing a natural Bourbon vanilla extract manifests itself in an unbalanced taste with phenolic, charcoal off-notes, lacking creaminess and mouthfeel, exhibiting no lingering sensation and displaying poor overall taste acceptability.

This chapter first reviews the chemical and physical factors contributing to the instability of vanilla flavor in low-fat ice cream and then presents a rigorous physicochemical mathematical model describing initial flavor performance. This nonequilibrium phase partition model accurately predicts the effects of fat on the vanilla flavor profile in ice cream. It also allows for the calculation of a reformulation factor for each chemical component of a compounded flavor to restore the original aroma and taste in a reduced fat product. Furthermore, the partition model enabled the design of a novel cryogenic fat enrobement technology for the vanilla extract since it cannot be reformulated. In this case we created a microenvironment for the flavor that mimics high fat ice cream.

Materials and Methods

Materials. All chemicals used in this study were of analytical grade. Food ingredients were purchased from suitable suppliers in Holland, Germany or the United States.

Determination of Partition Coefficients. Both water-to-air and oil-to-air partition coefficients were determined at 25°C using capillary tubes packed with XAD-4 beads according to the method of Etzweiler et al. (*5*).

Determination of Lactoperoxidase. Milk was subjected to various heat treatments and then used for the preparation of vanillin-containing custard. Residual lactoperoxidase activity in the custard was quantitated by determining the enzymatic oxidative conversion of added vanillin to vanillic acid. The custard (1.0 mL) was diluted with water (5.0 mL) and homogenized in an ultrasonic bath. Vanillin and vanillic acid were extracted with acetonitrile, separated from the precipitate by subsequent centrifugation and filtration, and analyzed by reverse-phase HPLC on a C18 column. The compounds were eluted with a solution of acetic acid and sodium

Table I. Ingredient Composition of Vanilla Ice Cream

Ingredient	Low-Fat Ice Cream (2% Fat)	High-Fat Ice Cream (15% Fat)
Heavy cream (36% fat)	1.40%	39.09%
Milk (2% fat)	82.10%	44.41%
Sugar	14.00%	14.00%
Na-CMC	2.00%	2.00%
Vanilla extract	0.25%	0.25%
Guar gum	0.10%	0.10%
Salt	0.08%	0.08%
Carrageenan	0.07%	0.07%

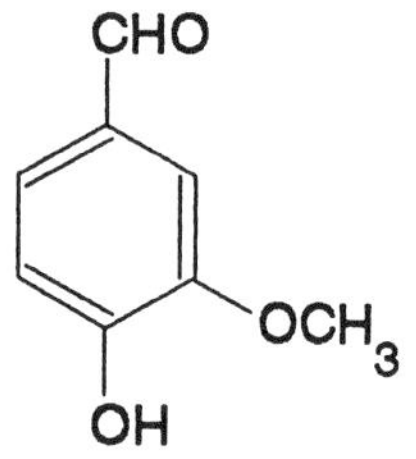

Figure 1. Structure of vanillin.

Table II. Enzymatic Oxidation of Vanillin

Heat Treatment of Milk	Lactoperoxidase Activity in Heat-Treated Milk	Vanillin Oxidation in Custard
75°C	6.1 U/ml	29%
80°C	<0.1 U/ml	11%
85°C	<0.1 U/ml	0%

phosphate in acetonitrile-water (20:80) and quantitated by measuring peak areas at both 254 and 275 nm.

Cryogenic Vanilla Enrobement. Vanilla extract was plated onto silicon dioxide and subsequently enrobed with a combination of a soft fat and an emulsifier using rotating disk encapsulation as described previously (*6*). The selection criteria for the enrobement oil include both its taste characteristics, nutritional properties and melting point range, i.e. it must be essentially molten slightly below body temperature in order to liquify and release the flavor during consumption of the ice cream.

Ice Cream Preparation. Full-fat and low-fat (2%) ice cream was prepared using an Ott freezer. The dry ingredients listed in Table I were pre-weighed and blended, and this dry mix was added to a cream and milk mixture. The new mixture was stirred for 3 minutes with a lightening mixer at high speed and allowed to set for 5 minutes. The mixture was restirred for 3 minutes and allowed to set overnight in a refrigerator. The next morning the product was restirred, poured into the hopper funnel of the Ott freezer, and mixed for 15 minutes. The flavor was added and blended into the ice cream immediately prior to placing the product into frozen storage. The level of vanilla extract was kept constant at 0.25% in all variables; the concentration of the enrobed vanilla was adjusted to achieve the same level of active materials.

The amount of overrun in all ice cream samples was maintained at 90.0% ± 5.0%.

Sensory Analysis. The organoleptic quality attributes of various vanilla flavored ice cream samples were compared by a professional taste panel consisting of 10 participants. A moderator first trained the panelists to establish sensory terms and then administered several qualitative descriptive analysis (QDA) tests. Overall flavor performance and acceptability in both low-fat and full-fat ice cream were also evaluated by 9-point hedonic tests.

Vanillin intensity in all ice cream samples stored in the freezer was monitored monthly over a period of 6 months by the same 10 trained panelists. Intensity values were scored on a scale of 1 to 15, 1 being the lowest.

Results and Conclusions

Deterioration of vanilla flavored ice cream during storage is a direct measure of vanillin instability. Vanillin (Figure 1), the main flavoring component of vanilla extract, may undergo the following types of reactions leading to its organoleptic loss:

1. Enzymatic Oxidation
2. Schiff Base Formation
3. Physical Interactions with Hydrocolloids

Vanillin easily forms the acetal in alcoholic solvents, but it is fairly stable to oxidation. Trace amounts of milk-derived lactoperoxidase, however, efficiently convert vanillin to vanillic acid as shown in Table II. Residual enzyme activity in

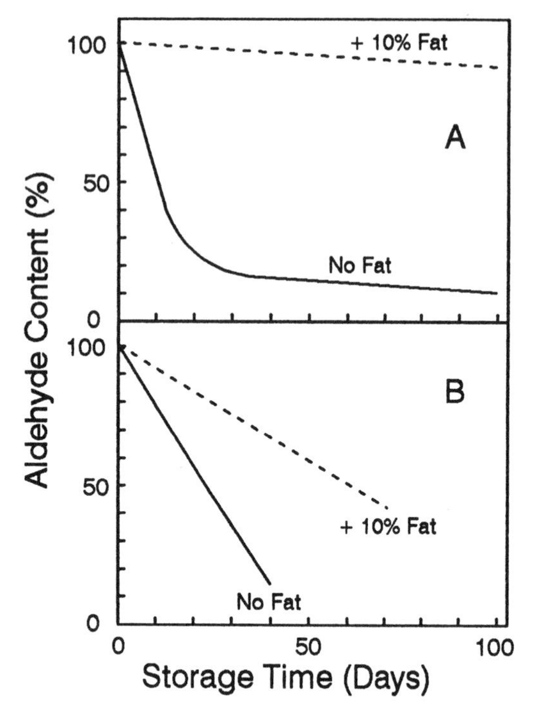

Figure 2. Protective effect of fat on the stability of aldehydes in the presence of 5% protein for: (A) 100 ppm citral, and (B) 100 ppm vanillin. Products were stored at 4 °C and analyzed periodically for citral and vanillin by combined extraction and UV spectroscopy.

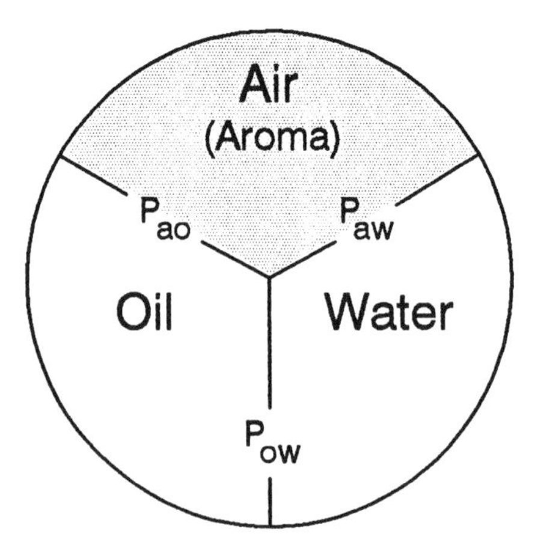

Figure 3. Schematic equilibrium 3-phase partition model.

milk pasteurized at 75 °C has been demonstrated to account for substantial vanillin loss in some European dairy desserts (J. P. Nelissen, unpublished results). Vanillin is completely stable in UHT-treated milk.

A considerable amount of vanillin loss during storage can be explained by a second type of chemical reaction involving other food ingredients, namely the condensation of the aldehyde function with primary amines. Schiff base formation occurs between vanillin and food proteins (*7*). Figure 2 illustrates how fat removal greatly accelerates this reaction and thereby shortens the shelf-life of low-fat or no-fat vanilla ice cream. An emulsion provides a distinct oil phase that can dissolve the aldehydic flavor. Decreasing or completely eliminating this protective phase drives the flavor into the aqueous medium in which the Schiff base condensation can take place. The shielding effect of even small amounts of oil increases proportionately with the hydrophobicity of the flavor chemical. The oil-to-water partition coefficient (P_{ow}) at 25 °C for vanillin (2.51) is approximately 100 times less than that for citral (240, neral; 309, geranial) which clearly accounts for the much more pronounced stabilizing effect of fat on citral than on vanillin in Figure 2. In the case of citral, a similar difference exists even between chemically pure citral and straight cold-pressed lemon oil, i.e. citral in the form of its natural raw material is substantially more stable than its pure chemical counterpart when used in a protein-containing aqueous solution (*7*).

The removal of fat in ice cream accelerates Schiff base condensation of vanillin with proteins by additional mechanisms. Fat replacement often results in an increase in total milk protein solids. At the same time proteins and vanillin may become further concentrated by pure water crystallization.

Most low-molecular weight organic molecules, particularly aldehydes like vanillin, are known to adsorb to food polymers. Many of these reversible physical equilibria have been studied extensively (*8-10*). Vanillin binds to most hydrocolloids through weak hydrophobic interaction. In addition, it forms strong phenolic hydrogen bonds with protein and also helical inclusion complexes with linear starches. Although vanillin reversibly bound to such hydrocolloids is still measurable analytically, it cannot be perceived organoleptically (*11,12*). The slow kinetics of flavor interactions with protein, starch and other polymers minimizes the extent of flavor desorption during mastication and swallowing of the food, which in turn precludes any pharyngeal volatilization necessary for aroma perception. Calorie reduction in ice cream exacerbates the gradual vanilla loss during storage due to this reversible adsorption to food hydrocolloids for similar reasons discussed above, including transfer of vanillin from oil to aqueous phase, freeze concentration of solutes, and replacement of fat with proteins, carbohydrates and gums.

While the above chemical and physical reactions of vanillin account for flavor loss in ice cream during storage, the initial flavor profile and balance are determined to a large extent by the phase partitioning behavior of the individual flavor chemicals. The effect of fat content on vanilla aroma perception can easily be estimated from the difference in the equilibrium headspace concentrations above low- and high-fat products.

The schematic diagram in Figure 3 illustrates the concept of an equilibrium 3-phase partition model. When a flavor compound is allowed to equilibrate between an emulsion and air in a closed system, the fraction of the compound present in the headspace is given by its mass balance (equation 1). From this mass balance we can easily derive the mathematical extraction equilibrium based on partition coefficients (equation 2) by substituting partition coefficients for flavor concentrations ($c_a = P_{aw}c_w$, $c_o = P_{ow}c_w$). In equation 2 either P_{ow} or P_{aw} can also be expressed in terms of P_{ao} according to equation 3, since a complete mathematical description of all three partition equilibria in a closed ternary system requires the measurement of only two partition coefficients. The partition coefficient P_{ow} is directly proportional to the hydrophobicity of the flavor compound, while P_{aw} and P_{ao} are a direct measure of its volatility in water and in oil, respectively.

$$f_{air} = \frac{c_a V_a}{c_a V_a + c_o V_o + c_w V_w} \qquad (1)$$

$$f_{air} = \frac{V_a P_{aw}}{V_a P_{aw} + V_o P_{ow} + V_w} \qquad (2)$$

$$\log P_{ow} = \log P_{aw} - \log P_{ao} \qquad (3)$$

where:

f_{air} = fraction of flavor released into air (headspace)
V_w = volume of water (mL)
V_o = volume of oil (mL)
V_a = volume of air (mL)
P_{ow} = oil-to-water partition coefficient ($[c]_o/[c]_w$)
P_{ao} = air-to-oil partition coefficient ($[c]_a/[c]_o$)
P_{aw} = air-to-water partition coefficient ($[c]_a/[c]_w$)
$[c]_w$ = flavor concentration in water (g/L)
$[c]_o$ = flavor concentration in oil (g/L)
$[c]_a$ = flavor concentration in air (g/L)

Equation 2 fails to consider the aroma release effects of proton dissociation of compounds like butyric acid. By substituting the Henderson-Hasselbach equation (*13*) in the mass balance for the total aqueous flavor concentration we derived equation 4. This modified equilibrium equation calculates the release of a flavor chemical into the headspace as a function of its pK_a and phase partition coefficients at any pH. Similarly, by including additional phase terms in the denominator of equation 1 the phase partition model can easily be expanded to calculate the effects of food hydrocolloids or packaging material on the flavor balance. In such 4-phase or even 5-phase partition equilibria the partition coefficients can be determined empirically or calculated from published affinity constants. This approximation reliably predicts flavor performance in complex systems, provided flavor levels are below the binding saturation limit. This latter assumption usually is correct due to the low flavor concentrations in the product.

$$f_{air} = \frac{V_a P_{aw}}{V_a P_{aw} + V_o P_{ow} + V_w (1 + 10^{pH - pK_a})} \quad (4)$$

where:

$pK_a = -\log K_a$
K_a = acid dissociation constant

The pertinence of equation 2 to a real food system was tested previously (*14*). Five flavor compounds of a wide range of volatilities and hydrophobicities were dissolved in whole milk in sealed glass containers. Flavor partitioning from this oil-in-water emulsion into the headspace was determined experimentally and calculated using equation 2. Excellent agreement between observed and calculated flavor release from milk was obtained.

The equilibrium phase partition model (equation 2) was also used to mathematically simulate the effects of fat removal from ice cream on three typical vanilla extract components. Figure 4 depicts the drastic changes in the static headspace composition as a function of minor oil reduction in the emulsion, particularly in the very low-fat region. Since the combined fractions of flavor chemicals in the air above a food are directly proportional to the perceived aroma, Figure 4 serves as an illustrative example to visualize the influence of fat reduction in ice cream on the overall flavor profile. Greatest havoc in flavor balance is wreaked by compositional changes in the very low-fat region. In our simplified vanilla extract example shown in Figure 4 reducing the fat content from 1% to 0% completely reverses the aroma profile. The disproportionate increase in the headspace concentration of ethyl caprylate (and many extremely hydrophobic aromatic compounds not shown in Figure 4) accounts for the typical vanilla imbalance in no-fat ice cream, containing strong phenolic, charcoal off-notes.

The nonideality of equation 2 due to minor equilibrium perturbations by opening a food container and removing some air in the process of smelling is a mathematical flaw in the equilibrium partition model of little practical consequence. However, under real food consumption conditions the effects of equilibrium perturbations can no longer be neglected. In order to apply the static equilibrium equation 2 to flavor perception during food consumption, the phase partition model was modified to include dynamic nonequilibrium boundary conditions. Recently de Roos and Wolswinkel (*14*) developed a physicochemical multiple extraction model to describe flavor release in the mouth. According to this model, solvent dilution with

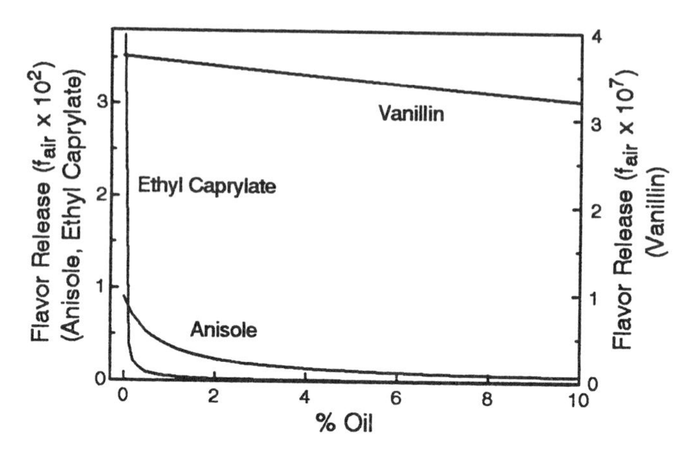

Figure 4. Effect of fat level on volatility of three vanilla flavor components. The fraction released into headspace was calculated using equation 2, assuming complete equilibrium in a closed system containing equal volumes of solution and air.

Table III. Correction Factors to Obtain Same Taste Intensity in 0% Fat as in 15% Fat

Ingredient	Log P_{OW}	Correction Factor
Vanillin	0.40	0.75
Phenol	0.89	0.50
p-Cresol	1.28	0.30
4-Ethylguaicol	1.74	0.13
Eugenol	1.99	0.08
Ethyl Benzoate	2.64	0.03
Methyl Cinnamate	2.79	0.02
Anethole	3.33	0.01

saliva, the partition coefficients of the flavor components and the resistance to mass transfer are the major factors determining the rate of transport of volatile flavor chemicals from the food into the vapor phase and then to the olfactory epithelium. Flavor dilution by saliva affects the overall 3-phase partition equilibrium and resulting aroma perception, the effect being most pronounced in a high fat food containing hydrophilic flavor components. In this case, the aroma perception by smelling is always much stronger than that by tasting and the overall flavor profile is also different.

Under nonequilibrium conditions the maximum headspace concentration predicted by equation 2 is never achieved. Therefore we adopted a multiple extraction model (*14*) in which a stream of air is constantly sweeping across the food and into the pharyngeal passage. The flavor is extracted consecutively from a small fraction of the food near the surface with infinitesimal volumes of fresh air. During each successive extraction phase equilibrium is achieved only at the product-to-air interface. The exact mathematical form of the expanded equation 2 varies with the food base, since the rate of phase equilibration depends largely on the matrix viscosity and other factors influencing the resistance to mass transfer. For each food category we empirically optimized a modified nonequilibrium version of equation 2 to provide the best fit between predicted and perceived flavor of reformulated matches for different fat variables. For example, in order to achieve the same aroma perception by mouth in a 15% and 0% ice cream base – excluding differences in temporal and mouthfeel sensations – we calculated the reformulation factors for eight different compounds present in vanilla extract (Table III). From these results it becomes apparent that there exists a wide range in reformulation factors for a single flavor. For some flavors containing both hydrophobic lactones and hydrophilic charged acids such reformulation factors may exceed a 1,000-fold range due to large differences in polarity. Compounded flavors can easily be reformulated for any new application. However, this broad spectrum of correction factors within a single flavor poses a severe challenge for reformulation of natural extracts. How can we adjust extraction conditions to achieve the high degree of specificity and control shown in Table III?

To solve this vanilla ice cream application problem we employed a type of reverse-phase engineering. Since we were unable to change the flavor composition, we decided to create a microenvironment surrounding the flavor that mimics the original full-fat base. We developed a cryogenic oil enrobement technology in which we plated the extract onto silicon dioxide and subsequently enrobed it in a combination of oil and an emulsifier using rotating disk encapsulation as described previously (*6*). An abbreviated summary of both hedonic and QDA sensory results is shown in Table IV. The enrobement not only restored most of the original flavor profile, but it also provided some lingering sensation and temporal sensory delay through including a plating support inside the capsules. At the same time the capsules provided partial immediate flavor release and left no objectionable waxiness or coating of the palate. These capsule features were designed by selecting a combination of emulsifier and enrobement oil with a melting point range slightly below body temperature in order to liquify and release the flavor during consumption of the ice cream.

Table IV. Sensory Attributes of Enrobed Vanilla Extract

Sensory Attribute	Free Vanilla in 15% Fat Ice Cream (Control)	Free Vanilla in 2% Fat Ice Cream	Enrobed Vanilla in 2% Fat Ice Cream
Vanilla Intensity	High	Medium	High
Off-Flavor	No	Yes	No
Flavor Lingering	Yes	No	Yes
Flavor Balance	Good	Poor	Good
Mouthfeel	Rich	Watery	Medium
Overall Acceptability	Good	Poor	Good
Similarity to Control		Low	High

Table V. Stability of Enrobed Vanilla Extract

Flavor System	Vanilla Intensity (Scale of 1 to 15)	
	Initial	3 Months
Free Vanilla Extract	6.8 +/- 1.8	5.3 +/- 1.1
Enrobed Vanilla Extract	8.7 +/- 2.9	8.0 +/- 1.5

Enrobement of vanilla extract not only improved initial flavor performance, but it also increased its storage stability in low-fat ice cream as shown in Table V. The limited mobility of vanilla flavor chemicals plated and trapped inside the frozen fat capsules significantly retards the chemical and physical interactions with food polymers. Therefore, the present approach simultaneously provided an identical solution to two independent applications problems.

In summary, fat and calorie reduction in vanilla ice cream reduces storage stability and distorts the initial flavor profile due to chemical reactions, interactions with food hydrocolloids and phase partitioning of flavor chemicals. Flavor balance can be restored by reformulating the flavor (in case of a compounded flavor) or by changing the flavor's microenvironment to imitate a full-fat base. The latter matrix mimicry was achieved using a novel cryogenic fat enrobement technology which also increased shelf-life stability.

The results from the present study revealed the major effects of food matrix composition on initial taste performance and storage stability of a flavor. A thorough fundamental understanding of these flavor-ingredient interactions provides the food developer with powerful tools to solve a number of flavor problems through simple formulation and processing changes. Also, the low-fat vanilla ice cream example illustrates the need for close collaboration between food processor and flavor house. Creative exploitation of the synergistic interactions between flavor and food is paramount to the design and systematic engineering of any successful product.

Acknowledgments

We thank J. A. Sarelse and E. McMillan for technical assistance and J. P. Nelissen for the measurement of lactoperoxidase-catalyzed oxidation of vanillin to vanillic acid.

Literature Cited

1. Glenn, C. *Food Engineering* **1994**, (5), 110-115.
2. Find/SVP. *FoodTrends™ '94,* sponsored by Thomas Food Industry Register and Find/SVP: New York, NY, 1994.
3. FMI/Prevention. *Eating in America: Perception and Reality. Shopping for Health 1994.* Food Marketing Institute/Prevention Magazine survey. Food Marketing Institute: Washington, DC, 1994.
4. Sloan, A.E. *Food Technology* **1994**, (7), 89-100.
5. Etzweiler, F.; Neuner-Jehle, N.; Senn, E. *Seifen, Oele, Fette, Wachse,* **1980,** *106,* 419-427.
6. Graf, E.; van Leersum, J. P. *U.S. patent,* **1994**, pending.
7. Gubler, B. A. In *Flavour 81;* Third Weurman Symposium, Schreier, P., Ed; de Gruyter: Berlin, Germany, 1981; pp 717-728.
8. Mills, O.E.; Solms, J. *Lebensm.-Wiss. Technol.,* **1984,** *17,* 331-335.
9. Solms, J.; Guggenbuehl, B. In *Flavour Science and Technology;* Sixth Weurman Symposium; Bessiere, Y.; Thomas, A. F., Eds.; John Wiley and Sons: Chichester, UK, 1990; pp 319-336.
10. Ng, P. K. W.; Hoehn, E.; Bushuk, W. *J. Food Sci.,* **1989,** *54,* 105-107.
11. Ng, P. K. W.; Hoehn, E.; Bushuk, W. *J. Food Sci.,* **1989,** *54,* 324-326.
12. Hansen, A. P.; Heinis, J. J. *J. Dairy Sci.,* **1991,** *74,* 2936-2940.
13. Lehninger, A. L. *Biochemistry;* Second Edition; Worth Publishers: New York, NY, 1975; p 50.
14. de Roos, K. B.; Wolswinkel, K. In *Trends in Flavour Research;* Maarse, H.; van der Heij, D. G., Eds.; Elsevier: Amsterdam, Holland, 1994; pp 15-32.

Chapter 4

Effect of Emulsion Structure on Flavor Release and Taste Perception

J. Bakker and D. J. Mela

Consumer Sciences Department, Institute of Food Research, Earley Gate, Whiteknights Road, Reading RG6 2EF, United Kingdom

In many foods the majority of the fat phase occurs as part of an emulsion, either oil-in-water (O/W) or water-in-oil (W/O). Several theoretical physico-chemical models of volatile flavor release have been developed, but there are no such models for tastant perception. In this chapter we present instrumental flavor release measurements from O/W and W/O emulsions of identical composition, by determining the headspace concentration as a function of time. These results indicate that theoretical models need to be further developed to predict flavor release. Sensory studies of simple taste compounds revealed a clear equality in perceived taste intensities of O/W and W/O emulsions, and it is suggested that this could be accounted for by phase reversion of W/O to O/W as a result of dilution with saliva in the mouth.

Flavor formulations are often specifically designed for foods with a particular level and composition of fat; hence, manipulations of the fat phase may markedly affect the perceived flavor characteristics of food products. In many foods the majority of the fat phase occurs as part of an emulsion, either oil-in-water (O/W), such as milk, or water-in-oil (W/O), such as butter. Despite considerable academic and industrial interest in fat modification of foods, there are few published studies addressing the role of fat content and processing on flavor release in general, or from emulsions specifically.

Physico-chemical Factors in Flavor Release

Several theoretical physico-chemical models of volatile flavor release have been developed (*1-4*), although they have not been fully tested in either instrumental or sensory experiments. Studies of the physico-chemical properties of volatile flavor compounds in model foods, such as their partition coefficients, can provide useful information regarding the possible effect of changing flavor compounds or the food matrix on the concentration of flavor in the headspace, as a function of the

0097–6156/96/0633–0036$15.00/0

concentration in the food matrix. The partition coefficient is often used as an indicator of flavor release from the food matrix, and the anticipated effect on sensory perception. Even 1% oil can affect the partition coefficients of aliphatic aldehydes, and the effect becomes more noticeable with increasing carbon number (*5*). Odor thresholds in vegetable oil determined for a series of aldehydes, ketones and pyrazines have been found to agree reasonably well with the calculated values derived from thresholds in aqueous solutions and the partition coefficients in water and in oil (*5*).

For sensory perception, the rate of flavor release is an important consideration, as it influences the time required before the threshold concentration of perception of a compound has been reached. De Roos and Wolswinkel (*6*) showed that the addition of fat to a volatile flavor solution alters the rate of flavor release as well as its partition coefficient. They reported an increase in flavor retention (relative to water) associated with the addition of 1% olive oil emulsified with 1% carboxymethylcellulose. The importance of the partition coefficient and fat content was also demonstrated in an evaluation of recognition threshold concentrations of styrene in O/W emulsions, which exhibited a linear increase with increasing fat content from 0.3 to 2.1 mg kg^{-1} (*7*). The authors attributed their finding to the very good solubility of styrene in fat, with the perception of styrene from O/W emulsions apparently determined by the concentration in the aqueous phase. The calculated concentration in the aqueous phase of the emulsion required for the flavor threshold recognition was determined to be fairly constant and in agreement with the threshold concentration.

Effects of Fat Phase Composition and Structure. Factors associated with the composition of the fat phase, such as the chain length and saturation of the fatty acid composition of oil, have been found to influence air-oil partition coefficients (*8*). Fat composition has also been shown to influence the release of flavor, as determined both by an instrumental method and by sensory flavor perception (*9*). In that work, the release and perception of diacetyl was studied in a slow melting saturated (stearin) fat with high solids content and also in a fast melting unsaturated (olein) fat. Both instrumental and sensory time intensity measurements showed that stearin gave a slower flavor release rate and a lower flavor intensity than olein fat.

With regard to emulsions, instrumental studies by McNulty and co-workers suggested an effect of both composition and structure on release of flavor from O/W emulsions (*1,2*). These authors formulated a flavor release model based on the assumption that movement of volatile, lipid-soluble flavors from the fat phase to the aqueous phase (a necessary step for subsequent release into the headspace) was induced by disturbing the equilibrium between the two phases, for example by dilution of the continuous aqueous phase. For O/W emulsions, instantaneous dilution with saliva in the mouth was assumed. The potential for flavor release in the mouth was said to be greater with (a) increasing distribution of the flavor in the fat phase, (b) an increasing fat fraction in the emulsion, and (c) increasing dilution; all of these factors which might be expected to generate the greatest potential for release of flavor from the fat to the aqueous phase on dilution. This model predicted a much slower perception of volatile flavors from W/O than O/W emulsions, since dilution of the continuous oil phase would not occur, and flavor transfer would therefore be very slow. It is not clear how this model might be applied to the release and perception of tastants, which tend to be hydrophilic materials dissolved in the aqueous phase.

Overbosch *et al.* (*4*) predicted that on theoretical grounds there should be no difference between the rates of release from O/W or W/O emulsions of the same oil fraction (ϕ=0.5). However, their own experimental data showed a clear effect of emulsion type, with the release rate of diacetyl into the headspace from an O/W emulsion being greater than from a W/O emulsion, both emulsions having the same oil fraction. This result was attributed to the different emulsifier systems used. Land (*10*) reported that, in two oil-water-air systems of identical overall composition, the unemulsified (presumably unhomogenized) system, had headspace concentrations different from the emulsified system. The effect of emulsification was different for the two flavors discussed, and it was suggested that this difference might relate to affinity for the interphase rather than the bulk phase. Haring (*11*) showed effects of an emulsifier (cetinol) on the release of a flavor compound using time intensity measurements. The presence of the emulsifier in an oil gave rise to the formation of smaller oil droplets as a function of mastication time. Perceived intensity of the flavor compound was increased and longer lasting in the presence of emulsifier.

In the two related experiments described here, we investigated the effect of emulsion type (O/W and W/O) on flavor release and tastant perception, both emulsions having the same oil fraction and prepared using the same emulsifier. The following studies were done: 1) release of a volatile flavor, as determined by instrumental measurements, and 2) tastant perception determined by sensory methods. In the former study, we determined the rate of release of diacetyl under conditions mimicking to an extent the events occurring in the mouth during eating (*12*). Since diacetyl is soluble in both the aqueous and the fat phase, partitioning between these phases and air was considered an important indicator of perception. In the taste experiment, we examined the perception of intensity of sucrose, sodium chloride, and citric acid from these emulsions (*13*). Since the tastants dissolved mainly in the aqueous phase, the breakdown of the W/O emulsion during eating was potentially of importance for the release and perception of tastants.

Materials and Methods

Emulsion Preparation. Both the O/W and W/O emulsions were prepared with equal amounts of sunflower oil and deionized water (ϕ=0.5). The same commercial sugar ester emulsifier (S-370 with HLB=3, Ryoto Sugar Esters, Mitsubishi Chemical Industries, Ltd., Tokyo, Japan) was used at 0.5% w/w to stabilize the systems for the volatile measurements and 1% was used in emulsions for sensory determinations. This emulsifier was selected for its very good emulsifying capacity in both types of emulsions and its neutral taste and odor. The sucrose stearate was mixed with a small, pre-warmed amount of the intended continuous phase (water or oil). The rest of the continuous phase material was added and the mixture stirred for 30 sec in a Silverton homogenizer (Vortmix, Hampton, Middlesex, UK) at 3000 rpm, followed by slow addition of the second phase (about 0.5 mL/sec) with continuous stirring. The emulsions were prepared using a small scale reverse flow microfluidizer (Model M-120E, Christianson Scientific Equipment Ltd., Gateshead, UK) and homogenized for 4 min at 300 bar pressure. Using these conditions, very stable emulsions were obtained, with a volume average droplet diameter of about 0.6 μ (*14*).

Flavor Release Samples and Measurements. A 5 mL sample of the emulsion at 25 °C was placed into a 40 mL vial. These were sealed with caps fitted with Mininert valves (Dynatech Precision Sampling Corporation; Baton Rouge, Louisiana, USA) and equilibrated at 25 °C for 15 min. Diacetyl (2 g/L) was injected into each vial and the headspace above this solution was sampled at regular time intervals to obtain a flavor release curve. Only one injection per vial was made. All experiments were carried out in triplicate.

The release of diacetyl from water into air was also determined by sampling the headspace while stirring in order to imitate solution movement in the mouth during eating. Flavor release was monitored from one vial also, while the contents were stirred with an average of 15 rotations per minute clockwise, followed by the same number anti-clockwise, in order to mimic eating conditions. After each sampling, the vial was very briefly vented to avoid creating a vacuum. Samples were taken at 5 min intervals for 90 min.

All headspace samples were analyzed using a gas chromatograph (5890, Hewlett-Packard Company; Palo Alto, CA) with a flame ionization detector and helium carrier gas at a flow rate of 5 mL/min (4 psi, equivalent to 28 kPa). Samples were injected in split mode (ratio 25:1) onto a 25 m x 0.53 mm i.d. WCOT fused silica CP WAX-52 CB capillary column with 1m retention gap (Chrompack Nederland B.V.; Middleburg, The Netherlands). The injector and detector temperatures were 150 °C and 250 °C, respectively, and the oven was maintained at 110 °C.

Sensory Samples, Tasting and Analysis. Five suprathreshold concentrations of sucrose (0.5, 1.0, 2.0, 3.0, 4.0 %w), NaCl (0.25, 0.35, 0.50, 0.70, 1.0 %w) and citric acid (0.15, 0.30, 0.50, 0.70, 1.0 %w) were each added to the pre-prepared O/W and W/O emulsions, producing 10 samples for each tastant. Samples were prepared approximately 18 hours before tasting, and held at 4 °C until use. Viscosities were assessed at 25 °C using a constant stress rheometer (Bohlin Rheology UK Ltd.; Cirencester, UK) fitted with a cone and plate measuring system, with viscosity determined at a shear rate of 40 s^{-1} and interpolated from the flow curve.

The samples were assessed by ten trained female sensory panelists (age 30-55 yr). All sensory assessments were done in three days, with two sessions per day separated by a 20-30 min break. All 10 samples of a single tastant were assessed in random order within a single session, and repeated in the second session that same day. Subjects were instructed to take a full plastic teaspoon (3 mL) of each sample from coded cups, evaluate them in the mouth for 2-3 sec and expectorate. Intensity of specific taste qualities and oral viscosity were evaluated on unstructured line scales anchored "nil" and "extreme" for sweetness, saltiness, and sourness (for sucrose, NaCl, and citric acid, respectively), and "thin" and "thick" for viscosity, using a computerized data collection system. Lukewarm mineral water was provided for rinsing between samples.

Sensory data were analyzed by ANOVA using a split-plot design with the sessions as blocks, the subjects (assessors) as whole-plot treatments, and the factorial

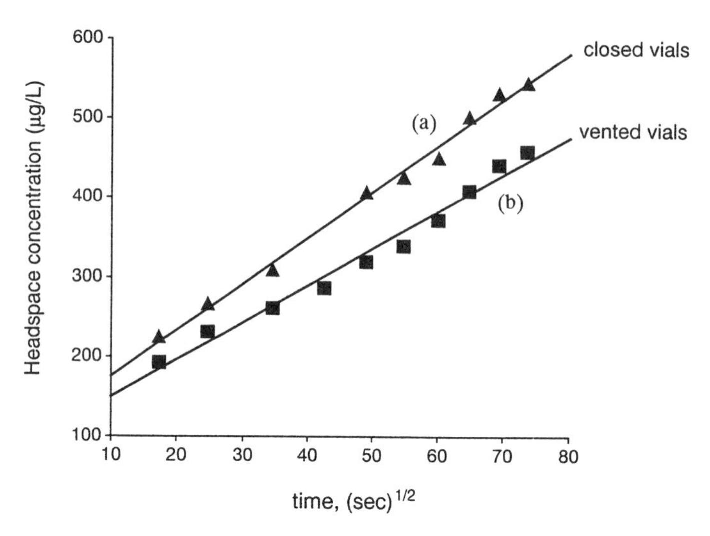

Figure 1. Release curves for diacetyl (2 g/L) from water at 25 °C of (a) closed vials, using one measurement per vial only and (b) three sets of measurements using one vial per set, briefly vented between measurements. Average experimental values (n = 39): slopes = 5.79 ± 0.15 and 4.65 ± 0.15 µg/l /s$^{1/2}$; intercepts = 117.9 ± 2.6 and 103.0 ± 2.6 µg/L respectively. (Adapted with permission from ref. 12. Copyright 1994, Elsevier Science Ltd., Kidlington, UK)

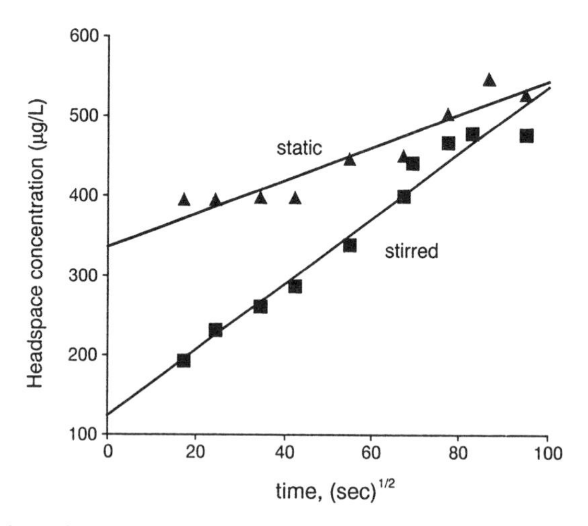

Figure 2. Release curves for diacetyl (2 g/L) from water at 25 °C under dynamic conditions using magnetic stirring and static conditions. Average experimental values (n = 39): slopes = 4.62 ± 0.16 and 2.17 ± 0.05 µg/L/s$^{1/2}$; intercepts = 124.3 ± 1.5 and 331.7 ± 1.5 µg/L respectively. (Adapted with permission from ref. 12. Copyright 1994, Elsevier Science Ltd., Kidlington, UK)

units (tastant concentration, emulsifier type) as sub-plot treatments. In addition, Pearson product-moment correlations and regression slopes were computed for physical and sensory measures.

Results and Discussion

Instrumental Flavor Release Measurements. Experimental release rates were determined by measuring the slope obtained by plotting headspace concentration against the square root of time. The rates quoted are for 35 mL headspace samples produced from a nominal 2 cm^2 surface area between the samples and the headspace. Figure 1 shows that the rate of release was greater when separate vials were used for the measurements, as determined by the difference in the slopes, rather than one vial vented after each measurement. The latter situation is likely to occur during eating, when one sample is kept in the mouth during the eating process, with regular venting when swallowing part of the sample or just breathing through the mouth. Under our experimental conditions small losses in headspace volatiles could occur as part of the sampling and venting procedure. Stirring of the sample, as might also occur during eating, increased the rate of release from water compared to release measured without stirring, as can be seen from differences between the slopes (Figure 2). Differences in concentrations on the intercept are due to the experimental procedure, and are not of interest here.

Figure 3 shows the release rates under static conditions for diacetyl from water compared with oil at 25 °C. The release rate during the linear part of the plot from oil was more than 5 times faster than from water. The time required to reach equilibrium was different for the two systems: equilibrium was established after only 15 minutes in the oil-air system, while it required 4 hours in the water-air system.

Figure 4 shows that the rate of release from the O/W emulsion was greater than from the W/O emulsion. Interestingly, release from both emulsions was rapid and faster than from the single phases. The release from the O/W emulsion was 1.5 times faster than from the W/O emulsion, as determined from differences between the slopes. This difference does not correspond with behavior predicted by the model of Overbosch *et al.* (*4*), which suggests a similar release rate for both types of emulsion. However, our data confirm their actual experimental results and showed that flavor release was twice as fast from O/W emulsions as from W/O emulsions. While these authors suggested that their finding may have been a consequence of using different emulsifiers for each emulsion, our samples were prepared with the same emulsifier for both O/W and W/O emulsions, suggesting this is a real effect, not an artifact.

The experimental rates of release from both emulsions were much higher than the calculated rates (*12*), presumably indicating that the model does not satisfactorily predict flavor release. One explanation for the difference between experimental and predicted release rates from the two different emulsions may be the increased interfacial surface for mass transfer. Additionally the structure of the interface formed between the droplets and the continuous phase may influence mass transfer between the phases of the two emulsions.

The difference between the release rate from water and from the O/W emulsion may be explained only in part by the relatively higher concentration of

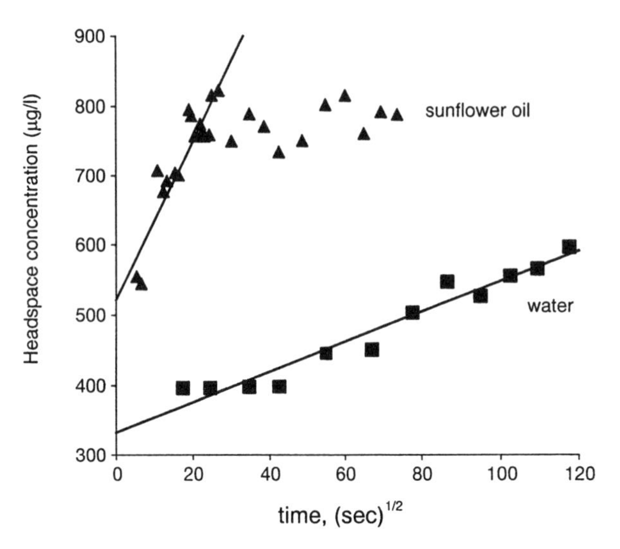

Figure 3. Release curves for diacetyl (2 g/L) from sunflower oil and water at 25 °C. For sunflower oil: average experimental values (n = 42): slope = 11.43 ± 0.33 µg/L/s$^{1/2}$; intercept = 522.1 ± 2.1 µg/L. Values for water are given in Figure 2. (Adapted with permission from ref. 12. Copyright 1994, Elsevier Science Ltd., Kidlington, UK)

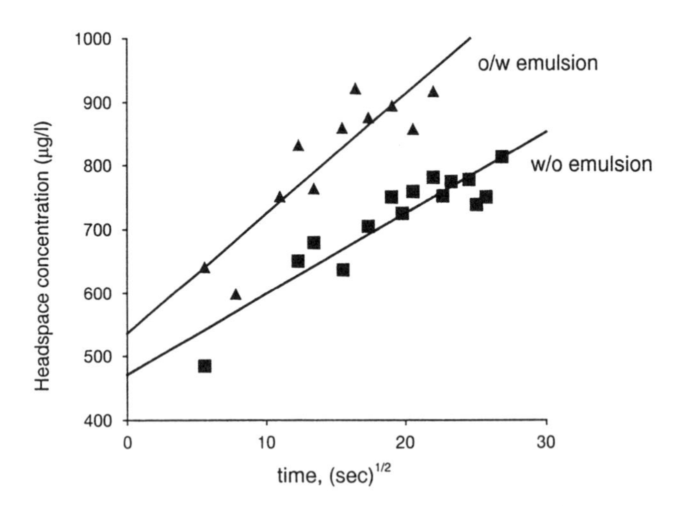

Figure 4. Initial release curves for diacetyl (2 g/L) from O/W and W/O emulsions (ϕ = 0.5 for both) at 25 °C. Average experimental values (n = 22): slopes = 18.87 ± 0.96 and 12.71 ± 0.40 µg/L/s$^{1/2}$; intercepts = 536.2 ± 4.8 and 470.8 ± 2.3 µg/L respectively. (Adapted with permission from ref. 12. Copyright 1994, Elsevier Science Ltd., Kidlington, UK)

diacetyl calculated to be present in the water phase of the emulsion. Factors such as increased available surface area and the dynamics of the emulsion may have to be taken into consideration. The present results clearly indicate that, besides the composition of the food, the structure also plays an important role and influences not only the rate of volatile flavor release, but also the amount released at equilibrium and hence the partition coefficient.

Sensory Perception of Tastants. There were no significant main effects of emulsion type on taste intensity for any of the three tastants (all $p > 0.05$), and this outcome is clearly illustrated in Figure 5. Similarly, slopes of concentration versus taste intensity were also not significantly different between O/W and W/O emulsions for any tastant.

In contrast to the taste data, consistent differences in perceived viscosity were apparent for NaCl and citric acid samples (main effect of emulsion type both $p = 0.001$), the O/W emulsions being perceived as thicker in both cases (*13*). While there are many reports describing interactions between taste and viscosity (*15-20*), these effects are probably related to the specific constituents and mechanical characteristics of the stimuli. In spite of the sensory results seen in the present study, there were in fact relatively small differences in measured viscosity between emulsion types, and this may explain why influences of viscosity on taste were not observed (*13*).

This study clearly indicates that emulsion type does not affect perceived taste intensity of sucrose, NaCl and citric acid within the range of component concentrations used here. The O/W and W/O emulsions have as their continuous phase distinctly different media, and the tastants used here are readily water soluble; thus, it is perhaps surprising to find relationships between tastant concentrations and intensities largely unaffected by emulsion type and associated differences in measured and perceived viscosity. Although it is always possible that statistically significant differences in taste intensity between emulsion types might be revealed by a larger panel or additional replications, the present data suggest that any such differences are likely to be small and of questionable practical significance.

Unfortunately, there are few other data which might be used to guide interpretation of these results. One possibility is that events within the mouth, particularly dilution with saliva, may substantially alter the characteristics of the samples. The models described by McNulty (*2*) highlight the potential role of saliva and sample dilution in the release of flavors from emulsions. Christensen (*21*) emphasized the active role of saliva in the perception of tastes and flavors, primarily for its diluting effect, but also for its powerful buffering capacity. She noted that the amount of saliva stimulated by any type of food is considerable, and the extent of its influence on taste perception process may depend on sample volume. In smaller samples, one would anticipate much more pronounced effects of saliva. In the case of a sample with lipid-continuous phase, there may be reversion into a water-continuous system (*9*). The standardized sample volume in this study was relatively small (3 mL), and conversion of the lipid-continuous phase of the W/O emulsions into a water-continuous sample therefore seems likely.

Thus, the very similar effects on taste intensity of O/W and W/O emulsions in the present study may result from both emulsion types sharing a common physical structure within in the mouth. This implies that taste intensity responses to these

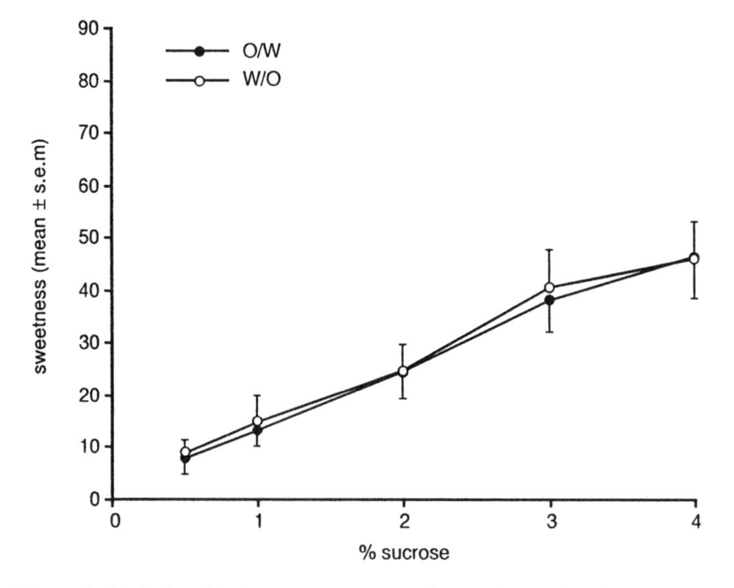

Figure 5. Relationship between concentration and perceived taste intensity of O/W and W/O emulsions containing sucrose, NaCl, or citric acid. (Reproduced with permission from ref. 13. Copyright 1995, Institute of Food Technologists, Chicago, IL.)

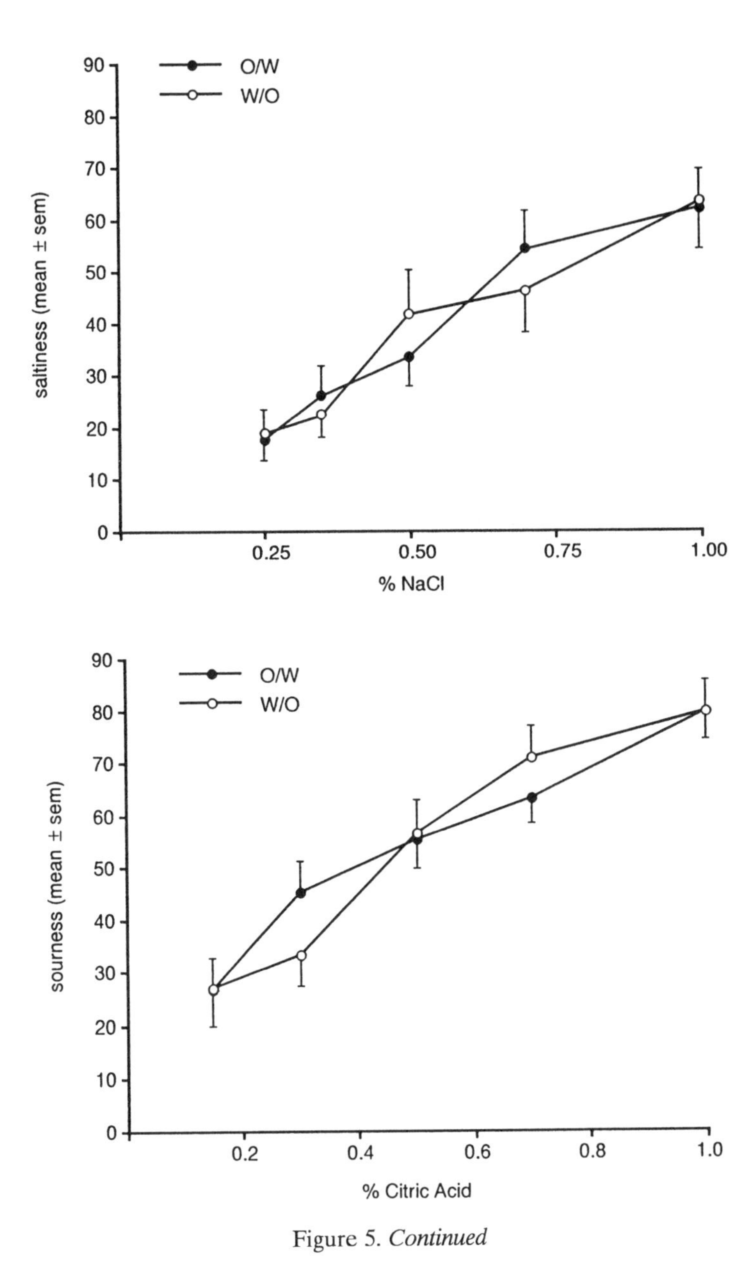

Figure 5. *Continued*

emulsion types could be different under conditions of larger sample size or altered salivary flow, and suggests that possible changes in sample structure within the mouth should be considered in predicting sensory responses. Another related consideration may be the effect of eating on the release of volatile flavors and tastants, and their subsequent perception. For example, mastication affects parameters such as available surface area for release of the taste and flavor compounds. Recent studies, focusing on the use of electromyography to assess chewing behavior, indicate consistent differences between subjects (*22,23*). These differences in chewing may influence the rate of breakdown of food structures, with possible consequences for the rate with which both flavor and tastants are released from the food matrix.

Conclusions

Instrumental measurements showed that the rate of release of diacetyl into air was faster from oil than from water. Conditions such as those occurring in the mouth, mimicked by stirring and headspace air changes, significantly affected the release rates. The emulsion structure also influenced flavor release: the rate of release from O/W emulsions was greater than from W/O emulsions, when both emulsions were prepared with the same emulsifier. These differences may be due to other effects, such as mass transfer rates between the interfaces. Further studies are needed to clarify this. In addition, the methodology used to measure flavor release needs to be developed, to allow on-line real-time flavor release measurements under controlled breakdown of food matrices. This would enable the empirical testing of theoretical models describing flavor release.

Sensory studies of simple taste compounds revealed a clear equality in perceived taste intensities of O/W and W/O emulsions, and it is suggested that this could be accounted for by phase reversion of W/O to O/W as a result of dilution with saliva in the mouth. Many flavor release and perception studies pay little attention to the potential influence or involvement of saliva (e.g., *4*), and generally assume that the same or similar physical systems exist outside and inside the oral cavity. The proposed explanation for our observations is speculative, and requires experimental confirmation.

Further elaboration upon the present results and the possible mechanisms involved may have important implications for understanding and predicting the sensory characteristics of a range of food emulsions.

Literature Cited

1. McNulty, P. B.; Karel, M. J. *Food Technol.* **1973,** *8*, 309-318.
2. McNulty, P. B. In *Food Structure and Behaviour*; Blanshard, J. V. M; Lillford, P., Eds.; Academic Press: London, 1987; pp. 245-258.
3. Darling, D. F.; Williams, D.; Yendle, P. In *Interactions of Food Components*; Birch, G. G.; Lindley, M. G., Eds.; Elsevier Applied Science: London, 1986; pp. 165-188.
4. Overbosch, P.; Afterof, W. G. M.; Haring, P. G. M. *Food Rev. Int.* **1991,** *7(2),* 137-184.

5. Buttery, W. E.; Guadagni, D. G.; Ling, L. C. *J. Agric. Food Chem.* **1973,** *21,* 198-201
6. de Roos, K. B.; Wolswinkel, K. In *Trends in Flavour Research*; Maarse H.; van der Heij, D. G., Eds.; Proceedings of the 7th Weurman Flavour Research Symposium; Developments in Food Science, vol. 35; Elsevier: Amsterdam, 1994; pp.15-32.
7. Linssen, J. P. H.; Janssens, A. L. G. M.; Reitsma, H. C. E.; Bredie, W. L. P.; Roozen, J. P. *J. Sci. Food and Agric.* **1993,** *61,* 457-462.
8. Maier, H. G. In *Aroma Research*; Maarse H.; Groenen P. J., Eds.; Pudock, Wageningen, Netherlands, 1975; pp. 143-157.
9. Lee, W. E.; Pangborn, R. M. *Food Tech.* **1986,** *40(11),* 71-78, 82.
10. Land, D. G. In *Progress in Flavour Research*; Land, D. G.; Nursten H. E.; Eds.; 2nd Weurman Flavour Research Symposium; Applied Science Publishers Ltd: Barking, UK, 1979; pp. 53-66.
11. Haring, P. G. M. In *Flavour Science and Technology*. Bessiere, Y.; Thomas, A. F., Eds.; John Wiley & Sons: Chichester, UK, 1990; pp. 351-354.
12. Salvador, D.; Bakker, J.; Langley, K. R.; Potjewijd, R.; Martin, A.; Elmore, J. S. *Food Qual. Pref.* **1994,** *5,* 103-107.
13. Barylko-Pikielna, N.; Martin, A.; Mela, D. J. *J. Food. Sci.* **1995**, *59*, 1318-1321.
14. Mela D. J.; Langley K. R.; Martin A. *Appetite* **1994,** *22,* 67-81.
15. Moskowitz, H. R.; Arabie, P. *J. Texture Stud.* **1970,** *1,* 502-510.
16. Pangborn, R. M.; Gibbs, Z. M.;Tassan, C. *J. Texture Stud.* **1978,** *9,* 415-436.
17. Christensen, C. M. *Percept. Psychophys.* **1980,** *28,* 347-353.
18. Kokini, J. L. *Food Tech.* **1985,** *39(11),* 86-92, 94.
19. Kokini, J. L. *J. Food Eng.* **1987,** *6,* 51-81.
20. Baines, Z. V.; Morris, E. R. *Food Hydrocolloids.* **1987,** *1,* 197-205.
21. Christensen, C. M. In *Interaction of the Chemical Senses with Nutrition*; Kare, M. R.; Brand, J. G., Eds.; Academic Press: New York, 1986; pp. 3-24.
22. Brown, W. E. *J. Text. Stud.* **1994,** *25,* 1-16.
23. Brown. W. E.; Shearn, M.; MacFie, H. J. H. *J. Text. Stud.* **1994,** *25,* 17-31.

Chapter 5

Interactions Between Lipids and the Maillard Reaction

L. J. Farmer

Food Science Division, Department of Agriculture for Northern Ireland and The Queen's University of Belfast, Newforge Lane, Belfast BT9 5PX, United Kingdom

The importance of lipid oxidation and the Maillard reaction for the generation of flavor in heated foods has been the subject of extensive research. However, these two reactions rarely occur in isolation and each may be expected to be modified by the reactants, intermediates and products of the other. Evidence is presented for the occurrence of such interactions in cooked foods; a number of long-chain heterocyclic compounds have been detected among the headspace volatiles from cooked meats and other foods. Aqueous model systems have been used to study the effect of lipid oxidation products on the volatile products of the Maillard reaction and its effect on the degradation of lipids. Phospholipids and triglycerides show different effects on the volatile products and aromas of the reaction between cysteine and ribose. The volatile thermal degradation products of lipids are also altered considerably in the presence of cysteine and ribose. Many of the effects of lipid-Maillard interactions appear to be due to reactions between carbonyl compounds (from the degradation of lipids or sugars) with amines (e.g., NH_3, amino acids, ethanolamine) or thiols (e.g., H_2S, mercaptoacetaldehyde).

Lipids have several important functions in the formation, release and perception of flavor. They are precursors of both desirable and undesirable odorous compounds formed by free radical oxidation reactions; they act as a solvent for many lipophilic odor compounds; their presence in foods affects the rate and period of flavor release in the mouth and their smooth texture contributes to mouthfeel and juiciness which can affect the perception of flavor *(1,2)*. However, lipids can also participate in the pathways of other flavor forming reactions. For instance, in the presence of amino acids and sugars, new volatile compounds are created and the products of both the Maillard reaction and the thermal oxidation of lipids are modified *(3-6)*.

0097–6156/96/0633–0048$15.00/0

Both lipid oxidation and the Maillard reaction are important reactions for the formation of flavor in many cooked foods and have been the subject of extensive research and many reviews *(e.g., 2,7)*. However, as most foods contain the precursors of both pathways, considerable interaction between them would be expected. This chapter describes some of the effects of interactions between the Maillard reaction and lipid oxidation pathways on the volatile aroma compounds.

Evidence for Lipid-Maillard Interactions in Foods

Certain compounds are characteristic of the interaction between the Maillard reaction and lipid oxidation. Long-chain heterocyclic compounds containing nitrogen or sulfur require for their formation the involvement of an amino acid and the participation of a fatty acid degradation product. A number of compounds of this nature have been detected in cooked foods; these include pyrazines, pyridines, pyrroles, oxazoles, thiazoles, thiophenes and trithiolanes with one or more butyl or longer alkyl substituent *(8)*. These types of compounds have been detected in fried foods (e.g., french fried potatoes and fried meat), where the oil may contribute to their formation, and also in baked potatoes and boiled meats. The presence of these compounds shows that, during the cooking of foods, the Maillard reaction interacts with lipids by several different mechanisms to give a variety of distinctive volatile products.

Table I shows that a large number of lipid-Maillard specific compounds have been detected in cooked meats. The routes of formation of these compounds illustrate the mechanisms by which lipid oxidation and the Maillard reaction can interact. For example, alkylthiophenes, alkylthiapyrans and alkylpyridines can all be formed by the reaction of H_2S or NH_3 with alkadienals *(4,9,10)*. Two of the classes of compounds listed in Table I have been identified more recently and have been subjected to further study to examine their routes of formation *(11,12)*.

Alkylformyldihydrothiophenes. Several 2-alkyl-3-formyldihydrothiophenes were detected in chicken, but none of the corresponding thiophenes were detectable *(12)*. However, both the alkylformyl-dihydrothiophenes and the alkylformylthiophenes were formed in aqueous model systems containing cysteine, ribose and phospholipid. This was possibly because of the higher temperature of the model systems (140 °C or 160 °C) compared with 100 °C for the meat.

The results of investigations in which the phospholipid was replaced by individual fatty acid methyl esters *(11)* showed that formation of these compounds depended on the fatty acid present. The higher molecular weight compounds were only formed in the presence of methyl oleate, indicating that the alkyl substituent originated from the omega end of the fatty acid chain. Indeed, further experiments showed that these compounds are the major products of the reaction between mercaptoacetaldehyde (from the Strecker degradation of cysteine) and 2-alkenals (from the thermal oxidation of unsaturated fatty acids).

Long-Chain Alkylthiazoles. The tentative identities of some of the alkylthiazoles detected in bovine heart muscle and *l. dorsi* and in chicken breast meat are given in

Table I. Occurrence of Lipid-Maillard Products in Cooked Meat

	Beef (skeletal muscle)	Beef (heart muscle)	Lamb	Chicken	Other meats[a]
1-Hexanethiol	*[b]			*	
1-Heptanethiol	*				
1-Octanethiol	*				
1-Nonanethiol	*				
Dipentylsulfide	*				
Methyloctylsulfide	*				
Methylnonylsulfide	*				
Dipentylsulfide	*				
Octylamine	*				
2-Butylpyridine			*	*	
2-Pentylpyridine	+,*(fr)[c]		*	+,*(fr)	*(tk)
3-Pentylpyridine			*		
5-Methyl-2-pentylpyridine			*		
2-Methyl-5-pentylpyridine			*		
5-Ethyl-2-pentylpyridine			*		
2-Ethyl-5-pentylpyridine			*		
3,5-Dibutyl-2-pentylpyridine				*(fr)	
2-Hexylpyridine			*		
2-Butylpyrrole	*				
1-Pentylpyrrole	*				
2-Butylpyrazine				*(fr)	
2-Methyl-3-butylpyrazine				*(fr)	
2-Butyl-3,5-dimethylpyrazine	*				
2-Butyloxazole					*(fb)
2-Methyl-5-pentyloxazole					*(fb)
4-Butyl-2,5-dimethyloxazole	*				*(fb)
5-Butyl-2,4-dimethyloxazole					*(fb)
2,5-Dimethyl-4-hexyloxazole	*				
2-Butylthiophene	*			*(fr)	
2-Pentylthiophene	*	+		*	
2-Hexylthiophene	*			+	*(tk)
2-Heptylthiophene	*				
2-Octylthiophene	*				
2-Pentylthiapyran	-	+		-	

Table I. (continued)

	Beef (skeletal muscle)	Beef (heart muscle)	Lamb	Chicken	Other meats[a]
2-Butanoylthiophene	*				
2-Heptanoylthiophene	*				
2-Octanoylthiophene	*				
2-Propyl-3-formyldihydrothiophene	-			+	
2-Butyl-3-formyldihydrothiophene	-			+	
2-Pentyl-3-formyldihydrothiophene	-			-	
2-Hexyl-3-formyldihydrothiophene	+			+	
3-Butyl-5-methyl-1,2,4-trithiolane				*(fr)	
3-Methyl-5-pentyl-1,2,4-trithiolane				*(fr)	
2-Butyl-4,5-dimethylthiazole	*			*(fr)	
4-Butyl-2,5-dimethylthiazole				*(fr)	
4,5-Dimethyl-2-pentylthiazole				*(fr)	
4,5-Diethyl-2-pentylthiazole	*				
2-Hexyl-4,5-dimethylthiazole				*(fr)	
2-Heptyl-4,5-dimethylthiazole				*(fr)	
4,5-Dimethyl-2-octylthiazole				*(fr)	
2-Butyl-5-ethyl-4-methylthiazole				*(fr)	
4-Ethyl-2-heptyl-5-methylthiazole				*(fr)	
5-Octyl-4-ethylthiazole[d]	tr	-		+	
5-Nonyl-4-ethylthiazole[d]	+	+		+	
5-Decyl-4-ethylthiazole[d]	tr	-		+	
2-Tridecyl-4,5-dimethylthiazole[d]	tr	+		-	
2-Tridecyl-4/5-ethylthiazole[d]	-	+		-	
2-Tetradecyl-4-methylthiazole[d]	-	+		-	
2-Tetradecyl-4,5-dimethylthiazole[d]	-	+		-	
2-Pentadecylthiazole[d]	tr?	tr		+	
2-Tetradecyl-4/5-ethylthiazole[d]	-	+		-	
2-Pentadecyl-4-methylthiazole[d]	+	+		+	
2-Pentadecyl-4/5-ethylthiazole[d]	+	+		+	

a Other meats: tk = turkey; fb = fried bacon
b * = reported in references *8* and/or *23*
- = not detected; + = detected; tr = trace
c fr = only reported in the fried meat; fat = only reported in the fat
d identification yet to be confirmed by synthesis

Table I. A number of analogous 2-alkylthiazoles have been synthesized by reacting 2-octanal with dicarbonyl compounds, H_2S and NH_3 *(12)* and the alkylthiazoles in Table I with a long chain in the 2-position are likely to be formed by this route *(13)*. However, these alkylthiazoles are unusual because they possess alkyl chains with 13 to 15 carbons and therefore would require a C14 to 16 aldehyde for their formation. Previously reported alkylthiazoles in fried chicken and french fried potatoes have possessed alkyl chains with only 6 to 8 carbon atoms *(14,15)*.

Cooked beef and chicken differ in the alkylthiazoles which are formed. In addition, the volatiles collected from heart muscle contain higher quantities and greater numbers of these compounds than either sirloin or chicken breast muscle *(12)*. This last observation suggests a possible origin for the long chain aldehydes required for the formation of these alkylthiazoles. Heart muscle has elevated levels of phospholipids compared with skeletal muscle and thus also contains more plasmalogen aldehydes (Table II; *16*). In addition, the most abundant plasmalogen aldehyde in bovine heart muscle is hexadecanal *(17)* which, by the mechanism of Takken *et. al.* *(13)* would give the pentadecyl substituted thiazoles, which were indeed prevalent. The occurrence of a range of plasmalogen aldehydes with 14 to 18 carbon atoms would explain the presence of the tridecyl, tetradecyl, and pentadecyl substituted thiazoles and also the detection of even higher molecular weight long-chain thiazoles which have yet to be identified (Farmer, unpublished data). Thus, evidence suggests that the very long chain aldehydes most probably originate from the plasmalogen aldehydes in the phospholipids.

Lipid-Maillard Interactions in Model Systems

Mottram and Edwards *(18)* investigated the effect of removing triglyceride or both triglyceride and phospholipid on the volatile compounds obtained from cooked beef and discovered that, in the latter case, there was a rise in the quantities of certain Maillard products; these results suggested that the presence of phospholipids could suppress the formation of Maillard products. A more detailed study of such effects required the use of model systems.

The effect of four lipids on the products of the Maillard reaction between cysteine and ribose, and of the Maillard reaction on the lipid degradation products, has been investigated using an aqueous model system *(5,6)*. Each model system contained cysteine and ribose alone, a lipid alone or Maillard reactants and lipid mixed together, as described previously *(5)*. Reactions were conducted in 0.5M phosphate buffer, pH 5.6 for 1h at 140 °C. Four lipids were investigated: beef triglyceride from beef adipose tissue (BTG), beef phospholipid from bovine muscle (BPL), egg phosphatidylcholine (PC) and egg phosphatidylethanolamine (PE). Phosphatidylcholine and phosphatidylethanolamine are the most abundant phospholipids in meat; the beef phospholipid contained about PC and PE in a ratio of approximately 2:1.

Lipid-Maillard Products. Long chain heterocyclic compounds detected in the model systems included 2-pentylpyridine, 2-alkylthiophenes, 2-alkylthiapyrans, 2-

Table II. Approximate Phospholipid and Plasmalogen Contents of Skeletal and Heart Muscle [a]

	Cattle			Sheep	
	Skeletal muscle	Heart muscle		Skeletal muscle	Heart muscle
Phospholipid content (% of wet tissue)	0.43-1.004[b]	2.4		1.24	2.75
Plasmalogen content (% of phospholipid fraction)[c]	26.9	28.5		8.9	31.4
Plasmalogen aldehydes (% of total aldehydes)[d]		PC	PE		
14:0		0.7	0.4		
14:0br		0.5	tr		
15:0		2.8	1.0		
15:0br		2.2	0.4		
16:0		62.0	28.8		
16:1		1.9	1.1		
17:0		2.5	3.2		
17:0br		8.0	3.7		
18:0		11.0	42.7		
18:1		7.0	18:1		

[a] Values were obtained from ref. 16 (and references cited within) and ref. 17.
[b] Values are from different references and may have been determined by different methods.
[c] Sum of contents of plasmalogen forms of phosphatidylcholine (PC) and phosphatidylethanolamine (PE).
[d] Aldehydes from monoacyl, alk-1-enyl forms of phosphatidylcholine and phosphatidylethanolamine.

(1-hexenyl)thiophene (*cis* and *trans*), 2-alkyl-3-formyldihydrothiophenes and 2-alkyl-3-formylthiophenes *(5,12)*. 1-Alkanethiols may also be regarded as compounds which require both lipid oxidation and the Maillard reaction for their formation. The formation of some of these compounds has been monitored in the four lipid-containing systems described above *(5)*. These compounds were not detected in the absence of lipid and only small quantities of some of them were detected in the triglyceride-containing system. The highest quantity of these compounds was detected in the model systems containing phospholipids and, among these, in the system containing PC *(5)*.

The higher amounts of long-chain heterocyclic compounds obtained in the phospholipids compared with the triglyceride is easily explained by the higher proportions of polyunsaturated fatty acids present; these would be expected to give increased quantities of unsaturated aldehydes by the well-documented free radical mechanisms *(2)*. In addition, the triglyceride was less miscible with the aqueous reactants and this would have inhibited interaction between the two reaction pathways. However, the differences in long-chain heterocyclic compounds and unsaturated aldehydes between PC and the other two phospholipids cannot be explained so simply; PC possesses fewer polyunsaturated fatty acids than PE or BPL and was also less miscible with the aqueous layer. The reason for the higher quantities of lipid-Maillard products obtained from PC compared with the other phospholipids is linked to the presence in PE and BPL of a free amine group, which can react with unsaturated aldehydes; this will be discussed later in this chapter.

A recent study *(19)* has shown that, compared with the aqueous systems discussed elsewhere in this chapter, far fewer lipid-Maillard specific products are formed in dry or low moisture model systems. It is suggested that, among other effects, water increases the mobility of the reactants, especially the phospholipid, thus facilitating lipid-Maillard interactions.

Changes in Maillard Products in Presence of Lipid. A variety of effects on the volatile products of the Maillard reaction between cysteine and ribose were observed when one of the four lipids was added to the system, depending on the class of volatile compound. Many were decreased by the presence of lipid while others were unchanged; some were reduced more by the phospholipids than the triglycerides while for others the converse was true *(5)*. Members of the same class of compounds tended to show the same effects; for instance, the formation of all the methyl-substituted thiazoles, 2-acetylthiazole, the pyrazines and the dithianones was unaffected by any of the lipids. The acylthiophenes, thienothiophenes, furanthiols, thiophenethiols and mercaptocarbonyl compounds were reduced by all the lipids; however, for the thienothiophenes and some acylthiophenes this reduction was greater in the presence of triglycerides than the phospholipids. The four mercaptocarbonyl compounds were reduced only slightly by the presence of beef triglyceride but were reduced considerably by all the phospholipids, while the furan- and thiophenethiols were reduced most by beef phospholipid and phosphatidylethanolamine.

In order to understand the different effects of the four lipids on the various Maillard products, it is necessary to consider their individual routes of formation. For example, the mercaptocarbonyl compounds can be formed by the reaction of dicarbonyl or α,β-unsaturated carbonyl compounds with H_2S *(20)*. It is likely that polyunsaturated fatty acids and their free radical degradation products compete with the sugar-derived carbonyl compounds for H_2S. The greater suppression of these compounds by phospholipids compared with triglyceride may be due to the low level of reactive polyunsaturated fatty acids in the triglyceride (less than 2% compared with 20% or more for the phospholipids).

The amounts of the furanthiols and thiophenethiols are reduced by all the lipids but especially by PE and BPL. These two lipids have two factors in common which may contribute to this effect. Both contain an ethanolamine amino group and both contain a high proportion of highly unsaturated fatty acids. The concentrations of ethanolamine groups are 20mM in egg PE, ca. 8mM in BPL compared with 41mM of cysteine amino groups *(5)*. Thus the ethanolamine amino groups could compete with the amino acid for reaction with carbonyl compounds, thus interfering with the Maillard pathways. The other factor is the concentration of polyunsaturated fatty acids; fatty acids with 2 or more double bonds make up 32% BPL, 34% PE compared with 22% of PC. The effect is even more pronounced when one considers fatty acids with 3 or more double bonds: 25% BPL, 20% PE compared with only 6% of PC *(5)*. It seems probable that polyunsaturated fatty acids and their thermal degradation products could compete with sugar degradation products and Maillard products for reactive precursors such as H_2S and NH_3 etc. Thus, the presence of certain phospholipids reduces the availability of reactive precursors for the Maillard pathways leading to the formation of furan- and thiophenethiols *(21)*.

Changes in the Products of Lipid Oxidation. Not only did the presence of lipids modify the production of typical Maillard products but the Maillard reaction also caused large changes to the aliphatic products of the thermal oxidation of lipids *(6)*. This effect was most pronounced for the aldehydes; the levels of all the aldehydes from all the four different lipids were reduced in the presence of cysteine and ribose (Table III). The saturated aldehydes were reduced by a factor of about two, while in most cases the unsaturated aldehydes were reduced by a factor of 10 or more.

Also evident from Table III is the large difference in amounts of unsaturated aldehydes formed from the four lipids; PC and even BTG give more of these compounds than the more unsaturated PE and BPL The probable cause of this difference is that both BPL and PE possess a free ethanolamine amino group which reacts with the aldehydes (or their precursors) to give Schiff's bases, thus removing them from the reaction systems and inhibiting the formation of some of the long-chain heterocyclic compounds. Similar reactions with the amino groups of cysteine are almost certainly responsible for the observed suppression of aldehyde formation in the presence of Maillard reactants.

It is uncertain why the addition of cysteine and ribose to BPL should cause less suppression of the unsaturated aldehydes than for the other lipids (Table III), as both PE and PC (the major constituents of BPL) showed large reductions in the amounts of these aldehydes formed.

The antioxidative effect of the Maillard reaction has been well documented *(22)*. However, the reduction in aldehydes by the addition of cysteine and ribose did not represent an overall antioxidative effect, as other lipid oxidation products (2-alkylfurans and 1-alcohols) were unchanged or even increased by the presence of cysteine and ribose (Table III); the mechanisms for some of these effects requires further investigation. Therefore, it is likely that the Maillard reaction reduces the quantities of aldehydes by suppression of individual pathways rather than of free radical oxidation as a whole.

Many unsaturated aldehydes possess low odor thresholds and these compounds contribute both to the desirable flavors of foods and to rancid off-

Table III. Quantities (ng) of Compound Classes Monitored in the Headspace of Lipids Heated Alone and in the Presence of Cysteine and Ribose *(6)*

Lipid:	BTG			BPL			PC			PE		
Cys + Rib:	-[a]	+	*Ratio*[b]	-	+	*Ratio*	-	+	*Ratio*	-	+	*Ratio*
n-Aldehydes (C6 - C10)	673	230	*0.34*	1381	514	*0.37*	1494	1060	*0.71*	3236	691	*0.21*
2-Alkenals (C5 - C11)	581	33	*0.06*	93.2	72.6	*0.78*	2602	66.9	*0.03*	67.6	0	*0*
2,4-Alkadienals (C7 - C11)	43.9	4.1	*0.09*	28.8	13.6	*0.47*	988	31.3	*0.03*	17.3	1.7	*0.10*
2-Alkanones (C6 - C10)	15.2	6.7	*0.44*	237	215	*0.91*	39.8	81.9	*2.1*	789	553	*0.70*
1-Alcohols (C6 - C9)	49.6	54.7	*1.1*	24.4	41.8	*1.7*	87.7	424	*4.8*	63.3	99.1	*1.6*
2-Alkylfurans (butyl - octyl)	4.7	14.9	*3.2*	272	230	*0.85*	154	372	*2.4*	327	495	*1.5*

[a] Lipids were heated in the absence (-) and presence (+) of cysteine and ribose

[b] Ratio = $\frac{\text{quantity of volatiles in presence of cysteine + ribose}}{\text{quantity of volatiles in absence of cysteine + ribose}}$

flavors. Thus, the role of the Maillard reaction in controlling the formation of such compounds may be important for the odor and flavor of many foods.

Conclusion

The studies summarized in this chapter provide evidence for a variety of interactions between the pathways involved in the thermal oxidation of lipids and those of the Maillard reaction. It is suggested that many of the observed effects of lipid-Maillard interactions are due to reactions between polyunsaturated fatty acids and their degradation products (especially unsaturated aldehydes) with reactive compounds such as amines (amino acids, NH_3), or thiols (H_2S, mercaptoacetaldehyde) and possibly also between ethanolamine groups and sugar-derived carbonyl compounds. Much work remains to be done to fully elucidate these mechanisms.

As many long-chain heterocyclic compounds do not possess strong odors, the main contribution of lipid-Maillard interactions to flavor is likely to be through the modification of the balance of odorous compounds, reducing the amounts of certain possibly potent aroma compounds and increasing the levels of others such that the overall aroma is that which we have come to expect of that food. Different lipids will modify the odor compounds formed in different ways, depending on their fatty acid composition, and the presence and nature of any plasmalogen aldehydes or polar groups. Thus the precise lipid composition of a food will influence the balance of the flavor-forming reactions and hence the overall aroma of the food.

Literature Cited

1. Forss, D. A. *Prog. Chem. Fats other Lipids.* **1972**, *13*, 181-258.
2. Chan, H. W.-S. *Autoxidation of Unsaturated Lipids*; Academic Press: London, 1987.
3. Whitfield, F. B.; Mottram, D. S.; Brock, S.; Puckey, D. J.; Salter, L. J. *J. Sci. Food Agric.* **1988**, *42*, 261-272.
4. Farmer, L. J.; Mottram, D. S.; Whitfield, F. B. *J. Sci. Food Agric.* **1989**, *49*, 347-368.
5. Farmer, L. J.; Mottram, D. S. *J. Sci. Food Agric.* **1990**, *53*, 505-525.
6. Farmer, L. J.; Mottram, D. S. *J Sci. Food Agric.* **1992**, *60*, 489-497.
7. Danehy, J. P. *Adv. Food Res.* **1986**, *30*, 77-138.
8. Whitfield, F. B. *Crit. Rev. Food Sci. Nutr.* **1992**, *31*, 1-58.
9. Hwang, S.-S.; Carlin, J. T.; Bao, Y.; Hartman, G. J.; Ho, C.-T. *J. Agric. Food Chem.* **1986**, *34*, 538-542.
10. Kawai, T.; Ishida, Y. *J. Agric. Food Chem.* **1987**, *37*, 1026-1031.
11. Farmer, L. J.; Whitfield F. B. In *Progress in Flavour Precursor Studies;* Schreier, P., Ed.; Allured Publ. Corp.: Carol Stream, IL, 1993; pp 387-390.
12. Farmer, L. J.; Mottram, D. S. In *Trends in Flavour Research;* Maarse, H.; Van der Heij, D. G., Eds.; Elsevier: Amsterdam, 1994; pp 313-326.
13. Takken, H. J.; van der Linde, L. M.; de Valois, P.J.; van Dort J. M.; Boelens, M. In *Phenolic, Sulfur and Nitrogen Compounds in Food Flavours*;

Charalambous, G. and Katz, I., Eds.; American Chemical Society: Washington DC, 1976; pp 114-121.
14. Tang, J.; Jin, Q. Z.; Shen, G.-H.; Ho C.-T.; Chang, S. S. *J. Agric. Food Chem.* **1983**, *31*, 1287-1292.
15. Ho. C.-T.; Carlin J. T. In *Flavor Chemistry Trends and Developments*; Teranishi, R.; Buttery R. G.; Shahidi F., Eds.; American Chemical Society: Washington DC, 1989, p 92.
16. Christie, W. W. *Prog. Lipid Res.* **1978**, *17*, 111-205.
17. Schmid, H. H. O.; Takahashi, T. *Biochim. Biophys. Acta.* **1968**, *164*, 141-147.
18. Mottram, D. S.; Edwards, R. A. *J. Sci. Food Agric.* **1983**, *34*, 517-522.
19. Mottram, D. S.; Whitfield, F. B. *J. Agric. Food Chem.* **1995**, *43*, 984-988.
20. Boelens, M.; van der Linde, L. M.; de Valois, P. J.; van Dort, J. M.; Takken, H. J. *Proc. Int. Symp. Aroma Research, Zeist*; Maarse, H.; Groenen, P.J., Eds.; Pudoc: Wageningen, 1975; pp 95-100.
21. van den Ouweland, G. A. M.; Peer, H. G. *J. Agric. Food Chem.* **1975**, *23*, 501-505.
22. Bailey, M. E.; Shin-Lee, S. Y.; Dupuy, H. P.; St. Angelo, A. J.; Vercellotti, J. R. In *Warmed-over Flavor of Meat*; St. Angelo, A.J.; Bailey, M.E. Eds.; Academic Press: London, 1987; pp 237-266.
23. Maarse, H.; Visscher, C. A. *Volatile Compounds in Food*; TNO-CIVO Food Analysis Institute: Zeist, The Netherlands (1989 and supplements).

Chapter 6

Flavor Binding by Food Proteins: An Overview

Timothy E. O'Neill

Department of Food Science and Technology, University of California, Davis, CA 95616

Perhaps the single most important criterion for consumer acceptance of foods is flavor. Proteins have little flavor of their own, but influence flavor perception via binding and/or adsorption of flavor compounds. Protein ingredients both transmit undesirable off-flavors to foods and reduce perceived impact of desirable flavors. This behavior is an important consideration in the design of food flavors, especially those intended for lowfat food formulations. Data from model systems illustrate that several factors determine the extent of interaction between proteins and food flavors, including the chemical nature of the flavor compound, temperature, ionic conditions and the structure and processing history of the food protein. Continued systematic study in this area will allow the optimal design of flavors for new formulated foods, elimination of transmitted off-flavors and development of efficient flavor carrier systems.

In a recent survey, ninety-one percent of shoppers ranked taste as an important factor in food selection (*1*). Much of the perceived flavor of foods is actually smell or aroma (2). The aroma response of foods is triggered by the volatile components of foods. Aroma is first perceived via the nasal cavities prior to placing food in the mouth (olfaction). Further aroma impressions are received via passage of volatile compounds through the pharynx at the back of the mouth to the nasal cavities as food is chewed (gustation). Volatile flavor compounds released in olfaction and gustation bind to or otherwise react with receptor proteins in the nasal passages to evoke a neural response. A response is evoked when a flavor compound reaches a critical threshold concentration in the nasal cavity or in the aqueous phase in the saliva.

0097–6156/96/0633–0059$15.00/0

Only very small concentrations of flavor compounds are generally necessary to elicit a sensory response, which ca be in the parts-per-billion range for some compounds. In addition, the difference between a pleasant or unpleasant flavor response can often be determined by small changes in flavor concentration. Due to the great sensitivity of the sensory organs to changes in flavor levels, interactions between flavor compounds and other components in foods can have profound affects on flavor perception by altering the rates of release volatile flavors. Flavor release from foods is a critical factor in flavor perception. Food lipids have the largest impact on perceived food flavors, by acting as solvents for lipophilic flavor molecules and reducing the rate of release of flavors into the air and aqueous phases when foods are being eaten (*3*). The overall effect of the presence of lipids in foods on flavor is a reduction in flavor intensity and a sustained release of flavors relative to non-fat food analogs (reviewed in reference *4*). Consumer demands for lowfat foods present a great technical challenge, since other components in foods much be engineered to duplicate the flavor partitioning behavior of the absent lipid component.

Proteins are particularly important in this regard, as many proteins have the ability to bind and sequester lipophilic molecules. This property was first observed to play an important role in the transmission of undesirable off-flavors in some products containing soy proteins (reviewed in reference *5*). The potential exists for the engineering of the food protein component for optimal flavor carrier and release properties. However, a fundamental knowledge of the phenomena and factors which affect flavor binding by proteins are necessary before this goal can be effectively realized. Two experimental approaches have provided the greatest insights thus far: headspace analysis and equilibrium binding measurements.

Headspace Analysis

Under the conditions normally existing in foods, flavor compounds are present only in extremely dilute concentrations and do not interact with one another. Hence, the partition coefficient can be defined via Henry's law as the ratio of the solute concentration in the vapor phase to its liquid phase concentration. The partition coefficient can be determined directly by measuring the equilibrium concentration of volatiles in the headspace above aqueous solutions by gas chromatography. Buttery et al. (*6*) demonstrated that the air-water partition coefficient depends on the chemical structure of the volatile compounds, generally varying according to chemical class and molecular weight in pure water systems (Table I). Use of the headspace analysis technique has illustrated the influence of other components in the aqueous phase on the volatility of flavor compounds. Lipids have by far the greatest effects, in many cases greatly reducing volatility (*3*). A number of researchers have shown that the addition of simple sugars and salts can affect the volatility of flavor compounds via their water structuring affects (*7-10*). Franzen and Kinsella (*11*) studied the binding of a homologous series of aldehydes by various food proteins in aqueous systems and found that the amount of bound flavor depends on the type, amount and composition of the protein tested, as well as the presence of lipids. Gremli (*12*) demonstrated that soy proteins reduce the volatility

Table I. Air-Water Partition Coefficients for Homologous Series of Aldehydes and Ketones at 25 °C

Compound	Partition Coefficient	Compound	Partition Coefficient
Acetone	1.6×10^{-3}	Propanal	3.0×10^{-3}
2-Butanone	1.9×10^{-3}	Butanal	4.7×10^{-3}
2-Pentanone	2.6×10^{-3}	Pentanal	6.0×10^{-3}
2-Heptanone	5.9×10^{-3}	Hexanal	8.7×10^{-3}
2-Octanone	7.7×10^{-3}	Heptanal	1.1×10^{-2}
2-Nonanone	1.5×10^{-2}	Octanal	2.1×10^{-2}
2-Undecanone	2.6×10^{-2}	Nonanal	3.0×10^{-2}
Acetaldehyde	2.7×10^{-3}		

SOURCE: Adapted from ref. 6.

of aldehydes and ketones by both reversible and irreversible binding phenomena.

Headspace analysis has played a key role in elucidating the impact of different food components on the volatility of flavor compounds. However, this method has several major experimental limitations. Headspace analysis lacks sensitivity and requires the use of large sample volumes for adequate detection, which can affect chromatographic analysis. The technique is also limited in its ability to distinguish and define the mechanisms through which interactions of flavors with food components occur.

Equilibrium Binding Phenomena

Flavor binding behavior can be characterized directly by equilibrium dialysis techniques. A semipermeable membrane is placed between two solutions in a twin chambered dialysis cell, one side containing a soluble protein and the other containing the flavor compound of interest, at subsaturating concentrations. The solutions are allowed to come to equilibrium and the concentration of the flavor compound is determined directly by extraction and gas chromatography. The most commonly used method for interpreting binding data makes use of the Scatchard equation (*13*). This model assumes the presence of a number of protein molecules, P, in solution, each possessing n indistinguishable and independent binding sites. The equilibrium between a ligand, L, and a protein with one binding site may be expressed as:

$$P + L = PL$$

The association constant, K, is defined via:

$$K = \frac{(PL)}{(P)(L)}$$

So:

$$(PL) = K(P)(L)$$

Since P(total) = PL + P,

$$(PL) = K(L)[P(total) - (PL)]$$

Or,

$$\frac{(PL)}{P(total)} = \frac{K(L)}{1 + K(L)}$$

Now,

$$\frac{(PL)}{P(total)} = \upsilon$$

where υ is the number of moles of ligand bound per mole of total protein. Thus:

$$\upsilon = \frac{K(L)}{1 + K(L)}$$

which is a statement of the law of mass action. If there are *n* independent binding sites, the equation for the extent of binding is simply *n* times that for a single site with the same intrinsic binding constant, *K*. Hence:

$$\upsilon = \frac{nK(L)}{1 + K(L)}$$

Or,

$$\upsilon/L = Kn - K\upsilon$$

Plotting υ/L vs υ gives the Scatchard plot, which has slope, $-K$, and y-intercept, nK. The above equation may be rearranged to give:

$$\frac{1}{\upsilon} = \frac{1}{nK(L)} + \frac{1}{n}$$

Plotting $1/\upsilon$ vs $1/L$ gives rise to the Klotz, or double reciprocal plot (*14*), which has slope, $1/nK$ and intercept, $1/n$. Experimental determination of the value of K as a function of temperature allows the determination of thermodynamic parameters related to binding. Thus:

$$\Delta G^0 = -RT \ln K$$

$$\Delta H^0 = -\frac{R\,d \ln K}{d(1/T)}$$

$$\Delta S^0 = \frac{\Delta H^0 - \Delta G^0}{T}$$

Model Systems

Equilibrium dialysis has been used to characterize the binding of a number of alkanones and aldehydes to bovine serum albumin (*15*), soy proteins (*16-18*) and bovine β-lactoglobulin (*19-20*). The case of bovine β-lactoglobulin is particularly interesting for food applications, as its functional and physical properties are extremely well characterized (*21*) and it is one of the few food proteins whose three dimensional structure has been fully determined (*22*). As the principal protein component of dairy whey, β-lactoglobulin has considerable use as a food ingredient and its flavor performance in foods is of considerable economic interest. Aside from the economic importance of this protein as a food ingredient, the example of β-lactoglobulin is particularly illustrative of the factors which affect flavor binding behavior by proteins.

Double reciprocal plots for binding of 2-heptanone, 2-octanone and 2-nonanone to β-lactoglobulin are shown in Figure 1. Within experimental error, the value obtained for the y-intercept is identical for all three compounds. This intercept indicates that there is one binding site for aliphatic ketones in native β-lactoglobulin B, assuming a molecular weight of 18,000 daltons for the protein. The slopes of these curves give the value of $1/nK$ from which the binding constants and corresponding free energy of association are readily obtained. These values, shown in Table II, indicate that the binding affinity of aliphatic ketones for β-lactoglobulin B increases proportionally with chain length. The free energy for the association of 2-alkanones with β-lactoglobulin changes by about -700 calories for 2-heptanone vs. 2-octanone, and by about -1000 calories from 2-octanone to 2-nonanone, suggesting that this association is primarily hydrophobic in nature.

The free energy changes corresponding to increase in chain length of one methylene group were -550 and -600 calories for the binding of these ketones to bovine serum albumin (*15*) and to soy protein (*16-17*), respectively. Robillard and Wishnia (*23*) reported that the increase in free energy of association was 1.07 Kcal per methylene residue for the binding of short alkanes to β-lactoglobulin A, consistent with the values found for alkanones.

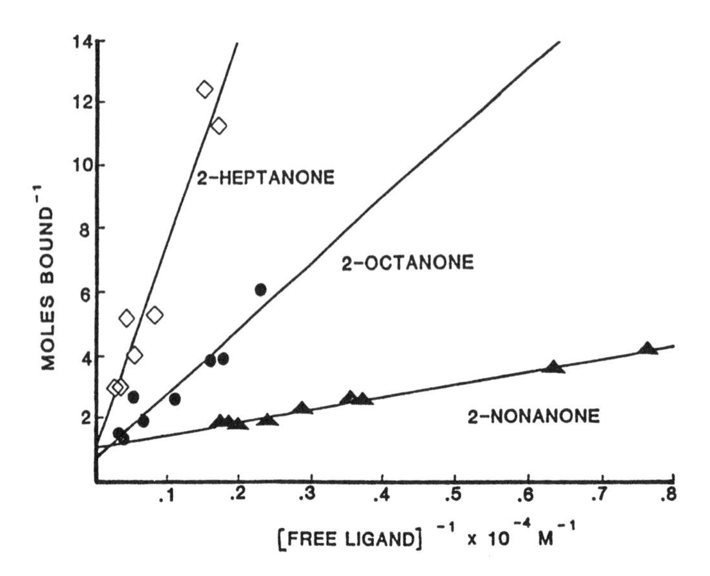

Figure 1. Effect of chain length on the binding of ketones to β-lactoglobulin B. Binding data are presented in the form of double reciprocal plots, using a molecular weight basis of 18,000 daltons for β-lactoglobulin B. (Reproduced with permission from ref. 19. Copyright 1987 American Chemical Society.)

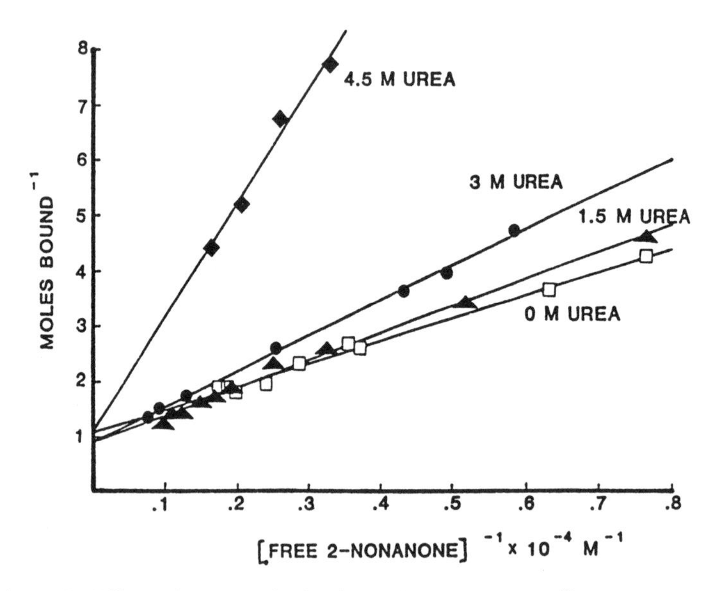

Figure 2. Effect of urea on the binding of 2-nonanone to β-lactoglobulin B. Binding data are presented as double reciprocal plots. (Reproduced with permission from ref. 19. Copyright 1987 American Chemical Society.)

Table II. Association Constants and Free Energy of Binding of 2-Alkanones to β-Lactoglobulin B

Ligand	K, M^{-1}	ΔG, kcal/mole
2-Heptanone	150	-2.98
2-Octanone	480	-3.66
2-Nonanone	2440	-4.62

SOURCE: Reprinted with permission from ref. 19.

The binding constants for the interaction of aliphatic ketones with β-lactoglobulin are higher than those obtained previously for either bovine serum albumin or soy protein. Damodaran and Kinsella (*16*) obtained a binding constant of 930 M^{-1} and 4 to 5 binding sites per molecule of protein for the interaction of 2-nonanone with soy protein, assuming a molecular weight of 100,000 daltons. On the same molecular weight basis, β-lactoglobulin would have approximately 5 binding sites and a binding affinity of 2440 M^{-1}, which is approximately two and one-half times that of soy protein. This suggests β-lactoglobulin may have a relatively large effect on the perceived flavors of foods when used as a food ingredient.

The 3-dimensional structure of β-lactoglobulin reveals that β-lactoglobulin has a hydrophobic core, which is accessible to the exterior of the protein molecule. It is this sheltered hydrophobic pocket which appears to be the sole binding site on the β-lactoglobulin molecule for a variety of nonpolar molecules, among them alkanes, sodium dodecyl sulfate, N-methyl-2-anilino-6-napthalenesulfonic acid as well as retinol and structurally related flavor compounds (reviewed in *19*, *24*). It seems likely that 2-nonanone also binds within this hydrophobic domain.

Effect of urea. Hydrophobic binding of ligands by proteins is dependent upon the existence of discrete hydrophobic regions within the protein molecule, which must be accessible to the ligand from the exterior of the protein molecule in order for the binding to occur. Protein denaturants, such as urea, can cause unfolding of the tertiary structures of proteins, thereby altering the structure of the available hydrophobic regions and the nature of hydrophobic binding. The effect of protein denaturation by urea on the binding of 2-nonanone by β-lactoglobulin B is shown in Figure 2. The slopes of the double-reciprocal plots increase with increase in urea concentration, corresponding to a decrease in binding affinity for 2-nonanone with increase in urea concentration. However, the number of binding sites for 2-nonanone remains constant at one throughout the tested range of urea concentrations. The results suggest that urea causes changes in the ligand binding domain, without disrupting it entirely.

X-ray crystallographic data suggest that a tryptophan residue resides in the interior of the β-lactoglobulin molecule, in the proximity of the most likely site for the binding of nonpolar molecules (*22*). Fugate and Song (*25*) determined that a

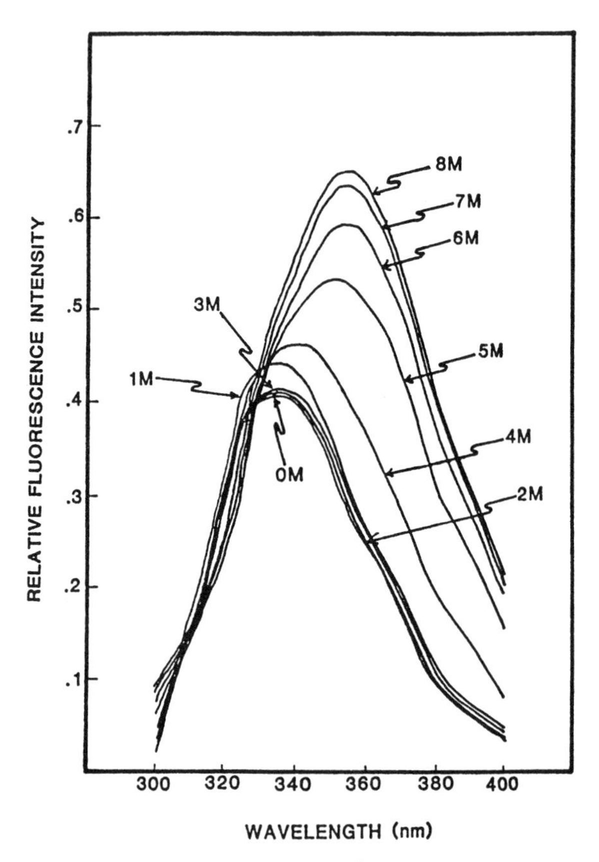

Figure 3. Fluorescence emission spectra of β-lactoglobulin B at various concentrations of urea in 20 mM phosphate buffer, pH 6.7. (Reproduced with permission from ref. 19. Copyright 1987 American Chemical Society.)

tryptophan residue is involved in the binding of retinol to β-lactoglobulin. This suggested that fluorescence spectra could be used to monitor conformational changes relevant to the environment of the hydrophobic binding site. The effect of urea upon the fluorescence spectra of β-lactoglobulin B is shown in Figure 3. Native β-lactoglobulin B exhibits maximum fluorescence intensity at 335 nm, consistent with tryptophan being the primary source of fluorescence emission. Little change occurs in the wavelength of maximum fluorescence intensity over the range from 0M to 3M urea concentration, whereas the maximum fluorescence intensity increases significantly from 4M to 8M urea concentration, with a shift in maximum fluorescence emission to a higher wavelength. This shift is consistent with the exposure of the tryptophan side chain to the aqueous solvent (*26*). These results suggest that the presence of urea may cause a partial unfolding of the β-lactoglobulin B molecule, partially exposing the single hydrophobic domain to the aqueous solvent and thus reducing the affinity of the protein for nonpolar molecules.

Effect of heat denaturation. Heat treatment is widely used in the food industry and can have profound effects upon the structure and functional properties of proteins (*27*). β-lactoglobulin is known to undergo a major change in its tertiary structure between 75-85 °C (*28-31*). The effect of heat treatment at 75 °C upon 2-nonanone binding is shown in Figure 4. The slopes of the double-reciprocal plots show an increase with increased heating time, corresponding to a decrease in the association constants, while the y-intercept decreases with heating time, indicating an increased number of binding sites. Thus, the nature of binding is changed by heat treatment. Fluorescence spectra (Figure 5) indicate an increase in maximum fluorescence intensity with increased heating time, with a small (2 to 4 nm) shift in wavelength of maximum fluorescence. This indicates that the tryptophan residues are slightly more exposed to the aqueous environment. Concomitantly, a conformational change has occurred, increasing the quantum yield of tryptophan. This fluorescence behavior could result from protein conformational changes, to protein aggregation or to both.

The occurrence of aggregation due to heat treatment of β-lactoglobulin B was confirmed by non-denaturing polyacrylamide gel electrophoresis of the heat treated samples. With increase in heating time at 75 °C the amount of monomeric β-lactoglobulin B decreased due to formation of aggregates. The results are presented graphically in Figure 6, expressed as percentage of native β-lactoglobulin remaining, plotted versus heating time. These data indicate that the conversion of the native structure into higher molecular weight aggregates is approximately second order with respect to heating time (not shown). β-Lactoglobulin B heated for 10 minutes and 20 minutes under these conditions shows approximately 60% and 80% conversion of monomeric structure into aggregates, respectively. Therefore, the 2-nonanone binding behavior is dominated by the presence of heat induced aggregates of the β-lactoglobulin monomer.

Hence, the nature and extent of the interaction between flavors and proteins may be modified by heat treatment. More specifically, proteins may bind more or

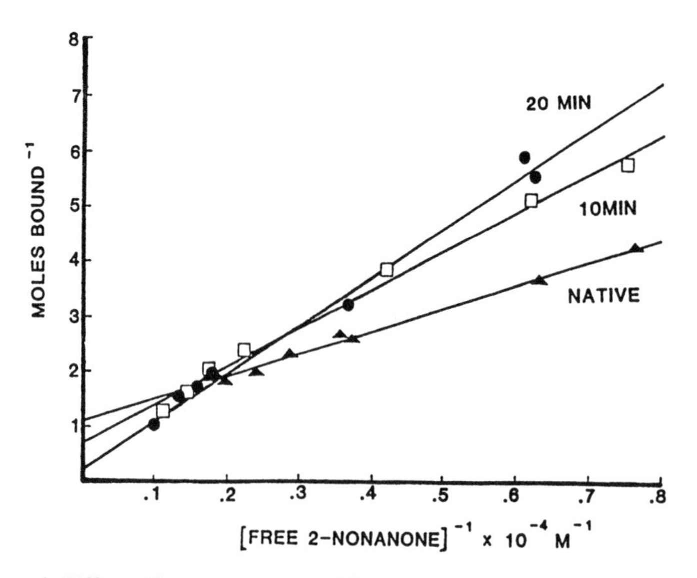

Figure 4. Effect of heat treatment at 75 °C upon binding of 2-nonanone to β-lactoglobulin B. (Reproduced with permission from ref. 20. Copyright 1988 Institute of Food Technologists.)

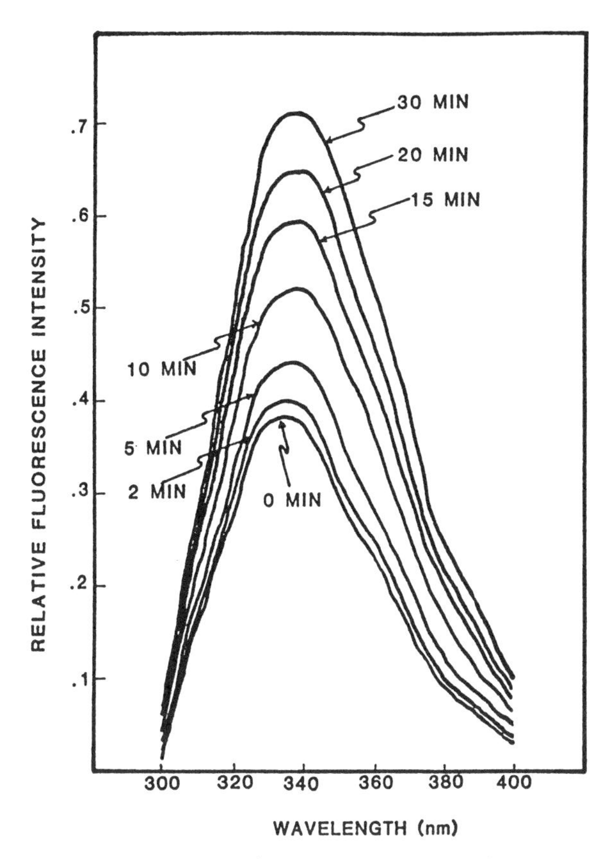

Figure 5. Fluorescence spectra of β-lactoglobulin B after heat treatment at 75 °C for various periods of time. (Reproduced with permission from ref. 20. Copyright 1988 Institute of Food Technologists.)

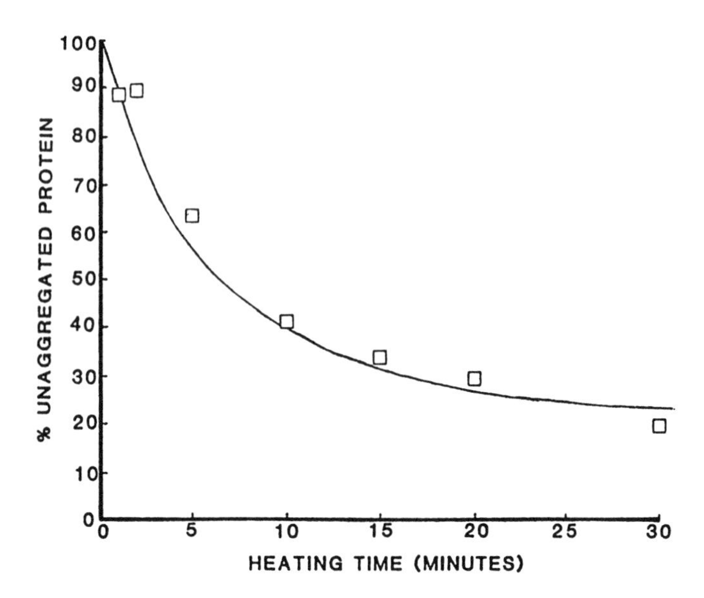

Figure 6. Effect of heating time at 75 °C on the amount of unaggregated β-lactoglobulin B remaining as determined from densitometer scans of electrophoretic gels of heated samples. (Reproduced with permission from ref. 20. Copyright 1988 Institute of Food Technologists.)

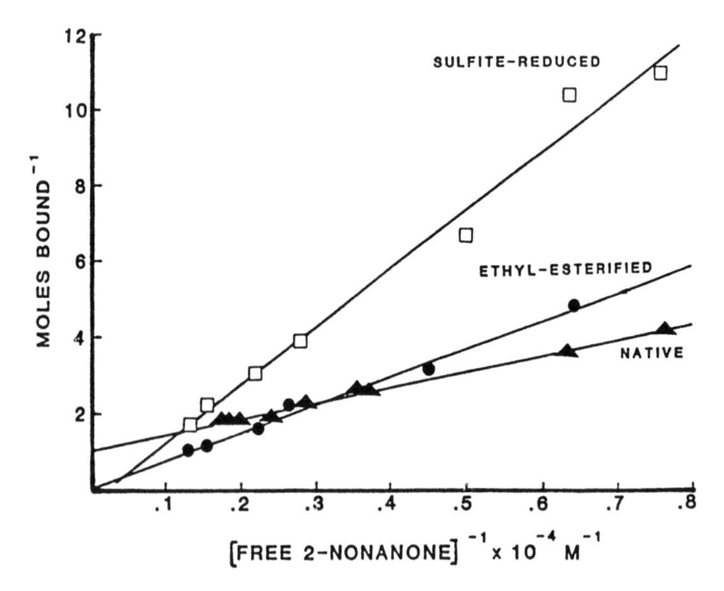

Figure 7. Effect of ethyl esterification or disulfide reduction on binding of 2-nonanone to β-lactoglobulin B. (Reproduced with permission from ref. 20. Copyright 1988 Institute of Food Technologists.)

less of a given flavor compound, depending upon the amount of thermal treatment. In this case, heat-denatured β-lactoglobulin will form aggregates which have an increased number of binding sites with somewhat lower binding affinity for 2-nonanone. Hence, the substitution of heat-treated β-lactoglobulin for native β-lactoglobulin in a flavored food might result in less binding of that flavor than allowed for in the original flavor formulation, resulting in a stronger or unbalanced flavor.

Effect of chemical modification. The effects of the chemical modification of β-lactoglobulin B by ethyl esterification and reduction of disulfide bonds with sodium sulfite on 2-nonanone binding are shown in the double-reciprocal plots in Figure 7. Both modifications resulted in dramatically different binding behavior from that of the native protein. The apparent binding of both modified proteins for 2-nonanone was markedly lower than that of the native protein. In addition, the y-intercepts of the double-reciprocal plots for the modified proteins are negative, suggestive of a very large number of weak binding sites for 2-nonanone within the modified protein molecules.

The optical density of native, ethyl-esterified and sulfite-reduced β-lactoglobulin B solutions (1%) at 600 nm were 0.007, 0.454, and 0.010, respectively. The high turbidity of the ethyl-esterified protein is likely to be caused by aggregation. This chemical modification of the carboxyl side chains of β-lactoglobulin B destabilizes the native conformation, causing substantial unfolding and aggregation. These aggregates, when dispersed in a solution exhibit different binding behavior since the binding site present in the native structure is no longer intact. The presence of large aggregates is likely to introduce heterogeneity of binding mechanisms into the system, since physical adsorption and entrapment may occur in addition to binding to hydrophobic sites.

The fluorescence spectra of native and sulfite-reduced β-lactoglobulin B are shown in Figure 8. This figure shows that the wavelength of maximum fluorescence emission is shifted from 335 nm for the native protein to 343 nm for the sulfite-reduced protein. In addition, reduction with sulfite results in a large increase in maximum fluorescence intensity. This indicates that reduction of the disulfide bonds in β-lactoglobulin results in a change in protein conformation, which affects the environment of the tryptophan residues. Thus, the modified protein has undergone a significant conformational change, wherein unfolding of the native structure has occurred, affecting the nature of 2-nonanone binding.

Sensory Analysis of Flavor Binding

Flavor binding by proteins has important implications on the perceived flavor impact in foods. Few sensory studies have been performed to confirm these effects directly. Ng et al. (*32-33*) observed a reduction in perceived vanillin flavor intensity in slurries of fababean protein micellar mass. Malcolmson et al. (*34*) demonstrated that the perceived intensity of chicken flavor is reduced by the presence of soy proteins in soup formulations. Hansen and Heinis (*35-36*) used trained taste panels to show that the perceived intensities of vanillin, benzaldehyde and *d*-limonene

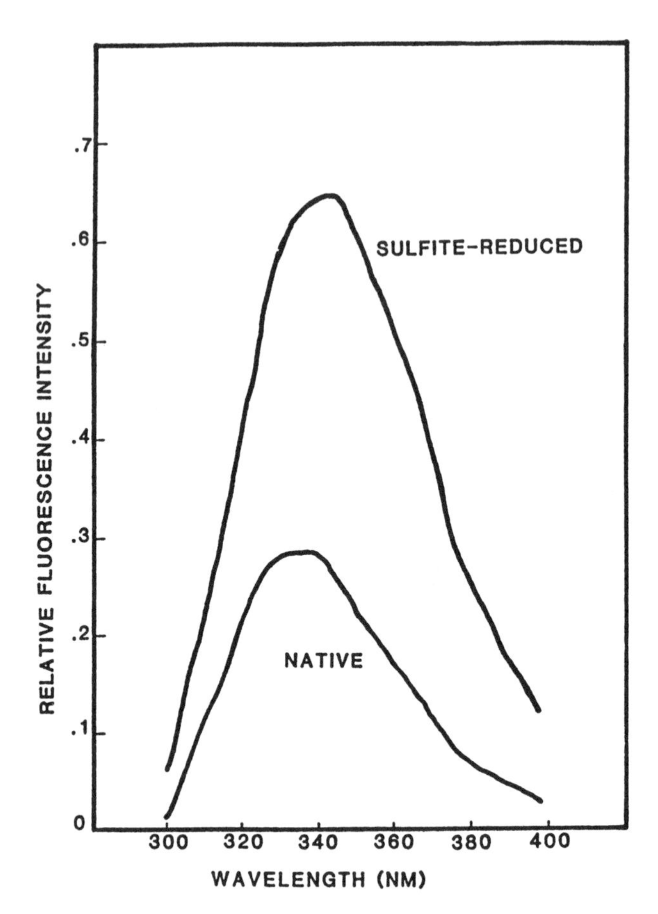

Figure 8. Fluorescence emission spectra of native and sulfite-reduced β-lactoglobulin B. (Reproduced with permission from ref. 20. Copyright 1988 Institute of Food Technologists.)

flavor are reduced in solutions containing either whey protein or sodium caseinate. The intensity of vanillin flavor perception was decreased by 50% in the presence of 0.5% whey protein concentrate. This is intriguing in light of the well characterized affinity of bovine β-lactoglobulin for lipophilic flavor molecules. However, it is not possible to distinguish from these experiments whether the decrease in perceived flavor intensity was due to reversible binding, or to irreversible binding (covalent attachment of flavor to protein via chemical reaction). Thus, the few controlled sensory studies performed to date support the concept that flavor binding by proteins may have significant effects on the perceived balance and intensity of flavors in foods. However, none of these studies have examined the mechanisms responsible for these effects. Since several different binding mechanisms could be responsible for reduction in flavor intensity in these model systems, it is imperative that an integrated approach be taken which includes both sensory and instrumental analytical methods.

Conclusions

Research in flavor chemistry over the years has successfully identified many of the key chemical components of food flavors, making it possible to assemble a wide variety of different, highly acceptable flavor formulations. These formulations are often complex mixtures of a number of compounds. The difference between acceptable and unacceptable flavor impact can depend on very small changes in the concentration or proportion of one or more components. Interactions of flavor compounds with food components provides a challenge to develop flavors for fabricated foods. Historically, this challenge has been met by laborious empirical practices in the hands of expert flavorists and sensory scientists. This is mostly due to a lack of detailed knowledge of the interactions of flavor compounds with food components.

Increased consumer demand for low-fat low calorie foods has led to the need for new types of flavor formulations. Lowfat food formulations present a huge challenge, as flavor formulations must be created which faithfully reproduce the flavor performance of the fat-containing food in a low-fat substrate. In these products the expected flavor balance, intensity and rate of flavor release normally associated with fat-containing foods must be reproduced. Satisfying these requirements on a case by case basis is time consuming and costly. An integrated, systematic research approach toward the understanding of flavor interactions with food components will be instrumental in the efficient design of new formulated foods. Such approaches should combine sensory and instrumental analyses to elucidate both the sensory impact and the mechanisms of flavor interactions with food components. These studies will eventually allow the more efficient design of both flavor formulations and food components which carry and release flavors optimally, providing the technical basis necessary for the food industry to rapidly accommodate changing consumer demands for nutritious and good-tasting foods.

Literature Cited

1. Anon., *1993 Supermarket Facts*; Food Marketing Institute: Washington, D. C., 1993.
2. Lawless, H. T. In "Encyclopedia of Food Technology," ed. Y. H. Hui., 2509-2525, Wiley, New York, 1992.
3. Kinsella, J. E. *Inform* **1990**, *1*, 215-226.
4. Hatchwell, L. C. *Food Technology* **1994**, February, 98-102.
5. MacLeod, G. and Ames, J. *Crit. Rev. Food Sci. Nutr.* **1988**, *27*, 219.
6. Buttery, R. G., Bomben, J. L., Guadagni, D. G. and Ling, L. C. *J. Agric. Food Chem.* **1971**, *19*, 1045.
7. Jennings, W. G. *J. Food Sci.* **1965**, *30*, 445.
8. Wientjes, A. G. *J. Food Sci.* **1968**, *33*, 1.
9. Nelson, P. E. and Hoff, J. E. *J. Food Sci.* **1968**, *33*, 479.
10. Nawar, W. W. *J. Agr. Food Chem.* **1971**, *19*, 1057.
11. Franzen, K. L. and Kinsella, J. E. *J. Agric. Food Chem.* **1974**, *22*, 675-678.

12. Gremli, H. A. *J. Am. Oil Chem. Soc.* **1974**, *51*, 95A-97A.
13. Scatchard, G. *Ann. Rev. N.Y. Acad. Sci.* **1949**, *51*, 660.
14. Klotz, I. M., Walker, F., Pivan, R. B. *J. Am. Chem. Soc.* **1946**, *68*, 1486.
15. Damodaran, S. and Kinsella, J. E. *J. Agric. Food Chem.* **1980**, *28*, 567-571.
16. Damodaran, S. and Kinsella, J. E. *J. Agric. Food Chem.* **1981**, *29*, 1249-1253.
17. Damodaran, S. and Kinsella, J. E. *J. Agric. Food Chem.* **1981**, *29*, 1253-1257.
18. O'Neill, T. E. and Kinsella, J. E. *J. Food Sci.* **1987**, *52*, 98-101.
19. O'Neill, T. E. and Kinsella, J. E. *J. Agric. Food Chem.* **1987**, *35*, 770-774.
20. O'Neill, T. E. and Kinsella, J. E. *J. Food Sci.* **1988**, *53*, 906-909.
21. Phillips, L. G., Whitehead, D. M. and Kinsella, J. *Structure-Function Properties of Food Proteins*; Academic Press: New York, 1994.
22. Sawyer, L., Papiz, M. Z., North, A. C. T. and Eliopoulous, E. E. *Biochem. Soc. Trans.* **1985**, *13*, 265.
23. Robillard, K. A. and Wishnia, A. *Biochemistry* **1972**, *11*, 3835.
24. Dufour, E. and Haertle, T. *J. Agric. Food Chem.* **1990**, *38*, 1691-1694.
25. Fugate, R. D. and Song, P.-S. *Biochim. Biophys. Acta* **1980**, *625*, 28.
26. Stryer, L. *Science* **1968**, *162*, 526.
27. Kinsella, J. E. *CRC Crit. Rev. Food Sci. Nutr.* **1976**, *7*, 219.
28. Hegg, P. *Acta Agric. Scand.* **1980**, *30*, 401.
29. de Wit, J. N. *Neth. Milk Dairy* **1981**, *35*, 47.
30. de Wit, J. N. and Klarenbeck, G. A. *J. Dairy Res.* **1981**, *48*, 293.
31. Park, K. H. and Lund, D. B. *J. Dairy Sci.* **1984**, *67*, 1699.
32. Ng, P. K. W., Hoehn, E. and Bushuk, W. J. *Food Sci.* **1989**, *54*, 105.
33. Ng, P. K. W., Hoehn, E. and Bushuk, W. *J. Food Sci.* **1989**, *54*, 324-346.
34. Malcolmson, L. J., McDaniel, M. R. and Hoehn, E. *Can. Inst. Food Sci. Technol.* J. **1987**, *20*, 229.
35. Hansen, A. P. and Heinis, J. J. *J. Dairy Sci.* **1991**, *74*, 2936-2940.
36. Hansen, A. P. and Heinis, J. J. *J. Dairy Sci.* **1992**, *75*, 1211-1215.

Chapter 7

Flavor Interaction with Casein and Whey Protein

A. P. Hansen and D. C. Booker

Department of Food Science, North Carolina State University, Raleigh, NC 27695

Processed milk protein-containing food products tend to retain less of the original perceived flavor as observed by sensory measurements. As the protein content of processed foods are increased to compensate for the reduction of fat, the potential exists for a corresponding reduction of flavor intensity due to flavor compound interactions with proteins. The purpose of this study was determine the extent of interaction between milk proteins and typical flavor compounds when the latter are mixed into ice cream during its manufacture. The model flavor compounds chosen for this study were vanillin, benzaldehyde, citral, and *d*-limonene. By fractionating the ice cream into fat, casein, and whey portions, one can determine the relative flavor concentration in each. Through quantitation of the amount of flavor in each fraction, the losses due to protein binding can be measured. The effect of these interactions upon sensory perception was also determined.

The acceptance of food products by the consumer is based on the sensory attributes of flavor, color, and texture. The aim of the food industry is to produce foods that are stable and have a good flavor and texture. One of the most important attributes of an acceptable food is the flavor as perceived by the consumer at the time of consumption.

The term "flavor" denotes the characteristics which stimulate taste, smell, thermal and tactile sensations. The flavor chemist is concerned with the compounds that contribute characteristic taste and aroma of foods. The common characteristics of food flavors are: (1) they consist of many components, some present in high proportions; (2) they exert their influence at extremely low levels; (3) they are highly specific with respect to molecular configuration; and (4) they tend to be volatile. Natural and artificial flavor systems contain a vast number of compounds

0097–6156/96/0633–0075$15.00/0

which contribute to the overall aroma and taste of a particular food. For example, chocolate and peanut flavors contain over 250 and 230 flavor compounds, respectively, some of which are more important than others to the characteristic flavor of the food. Simpler flavor systems, such as imitation vanilla and cherry flavors, also contain a variety of flavor compounds; however, each contains a sensory-dominating flavor component (vanillin and benzaldehyde, respectively) that contributes the characteristic taste and aroma.

Vanillin (**1**) is the principal component of vanilla extract, which is used widely by the food industry as a flavoring agent, specifically in confectionery products and beverages (*1*). Benzaldehyde (**2**) is a compound present in almond, cherry, and cinnamon-type oils, as well as in the essential oils of many flowers (*2*).

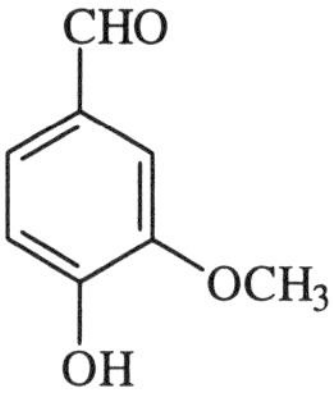

CHO OCH$_3$ OH

1

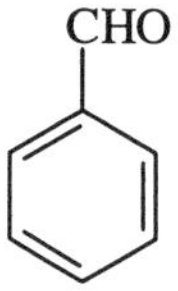

CHO

2

Citral has a strong lemon-like odor with a bittersweet taste. Commercially the product is a mixture of *cis*- and *trans*-isomers, geranial (**3a**) and neral (**3b**). *d*-Limonene (**4**) is one of the most widespread terpenes found in citrus peels. It is often used in frozen dairy desserts to produce a pleasant lemon-like odor.

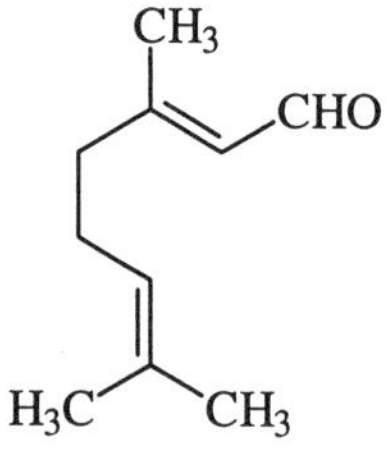

CH$_3$ CHO H$_3$C CH$_3$

3a

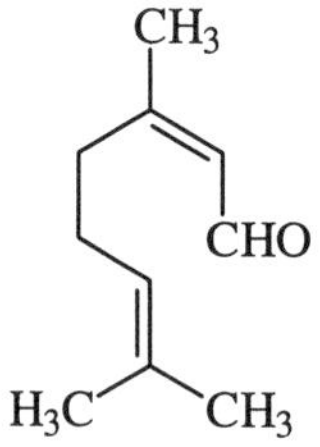

CH$_3$ CHO H$_3$C CH$_3$

3b

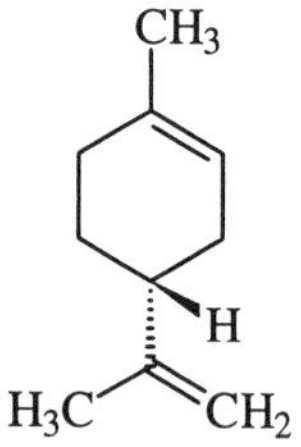

CH$_3$ H H$_3$C CH$_2$

4

Flavor Loss

Small changes in the levels of flavor compounds in food products can alter their sensory properties and render the flavor of the product unacceptable to the consumer. Flavor changes in food products have been attributed to several factors including light, processing conditions, ingredients, and packaging materials. Flavor loss can occur due to interactions between flavor compounds and other food ingredients. Protein-containing food products that are processed at high temperatures tend to retain less of the original perceived flavor (*3*). Hansen and Heinis (*3*) reported that vanillin flavor intensity, as measured by a 12-member trained taste panel, declined from 0.32 (moderately less than reference) to 0.15 (much less than reference) as the whey protein concentrate (WPC) level increased from 0.12% to 0.5% in flavored protein solutions. They later reported similar losses of benzaldehyde and *d*-limonene as WPC levels in flavored protein solutions increased from zero to 0.5% (*4*). Milk proteins are often added to lowfat frozen dairy desserts to impart smoothness and help to prevent weak body and coarse texture. Since 1988, protein-based fat substitutes have been available which simulate the mouthfeel of fat (*5*). In August 1991, a new version of a protein-based fat substitute was introduced which contained 100% whey protein. As the protein content of food is increased to compensate for the reduction of fat, the potential exists for reduction of flavor intensity due to flavor compound interactions with proteins. Even a small degree of interaction between flavor compounds and ingredients or packaging materials can reduce the amount available for sensory perception. Numerous studies have been conducted on the interaction of flavor compounds with β-lactoglobulin (β-lg) (*6-9*). β-Lactoglobulin readily binds certain alkanes, 2-alkanones, free fatty acids, triglycerides, and aromatic hydrocarbons (*6-9*). In addition to the primary binding site, β-lg is thought to contain other hydrophobic areas capable of undergoing interactions with apolar molecules (*7-10*).

The purpose of this study was determine the extent of interaction between milk proteins and typical flavor compounds when the latter are mixed into ice cream during its manufacture. The model flavor compounds chosen for this study were vanillin, benzaldehyde, citral, and *d*-limonene. By fractionating the ice cream into fat, casein, and whey portions, one can determine the relative flavor concentration in each. Through flavor quantitation of each fraction, losses due to binding with the protein can be measured. The effect of these interactions upon sensory perception was also determined. Perceptual and chemical data were determined for benzaldehyde and vanillin, while only perceptual differences were obtained for citral and *d*-limonene.

Materials and Methods

Materials. Sodium caseinate (CAS) and whey protein concentrate (WPC) were obtained from New Zealand Milk Products (Petaluma, Ca). Proximate analyses furnished by the manufacturer indicated that CAS (Alanate 180) contained 91.1% protein, 3.5% ash, 4.0% moisture, 1.1% fat, and 0.1% lactose and gave a pH of 6.6 in 5% aqueous solution at 21°C. The WPC (Alacen 855) contained 76.5% protein, 3.5% ash, 4% moisture, 3.5% fat, and 12.5% lactose and gave a pH of 6.7 in 5% aqueous solution at 20°C. Food-grade vanillin was obtained from Rhone Poulenc, Princeton, NJ. Food-grade benzaldehyde, *d*-limonene, and citral were obtained from Mother Murphy's Flavors, Greensboro, NC.

Determination of Vanillin in Ice Cream Mix Fractions. Thirty pounds of ice cream mix containing 10% milk fat and 8.5% milk solids non-fat (MSNF) were prepared using the formulation shown in Table I, pasteurized at 165° F for 30 min., homogenized at 1500 psi first stage, and 500 psi second stage and refrigerated for 24 hours.

Table I. Ice Cream Mix Formulation

Ingredient	*Amount*
33% Cream	9.09 lbs.
Milk Powder	1.83 lbs.
Cane Sugar	4.50 lbs.
Fine Guar Gum	27.2 grams
Water	14.52 lbs.

One mL of a 50-mM solution of vanillin in absolute ethanol was then added to 500 mL of ice cream mix and stirred on a magnetic stirrer for 1 hour. Three 50-mL portions of the flavored mix were centrifuged at 30,000 rpm for 270 min. at 15° C to yield three fractions designated as fat (top layer), whey (liquid), and casein (protein precipitate). Each fraction was weighed and extracted four times with 100 mL of methylene chloride. The last extraction was allowed to sit overnight in the separatory funnel at 15 °C. The four extractions of each fraction were pooled and refrigerated for 3 hours. The extracts from the fat layer were cold-filtered through Whatman #1 filter paper to remove solidified fat particles. The whey and casein extractions were concentrated in a rotary evaporator to 5 mL, transferred to a conical test tube and dried under a stream of nitrogen gas to a volume of 1 mL.

After concentration of the fat extract to 5 mL in a rotary evaporator, the contents were transferred to a conical test tube and completely dried under a stream of nitrogen gas, resulting in fat residue. The remaining residue was then increased to 25 mL with 100% ethanol and centrifuged at 20,000 rpm for 90 min. at 15 °C, yielding a fat and an ethanol layer. The ethanol layer was removed and the fat layer was again extracted with 25 mL of ethanol. The two ethanol extractions were combined and concentrated to 5 mL in a rotary evaporator. The 5-mL sample was then transferred to conical test tubes and dried under nitrogen gas, then raised to 1 mL with methylene chloride. The additional extraction of the fat fraction with ethanol was necessary to remove fat solubles initially extracted with methylene chloride.

A Shimadzu Mini-2 gas chromatograph, equipped with a 50 M Econocap SE-54 capillary column and a flame ionization detector (FID) was used to quantify vanillin in the samples. Flow rates were as follows: helium (carrier) 2.5 mL/min.; FID detector: hydrogen 35 mL/min., air 180 mL/min. The column was held for 12 minutes at an initial temperature of 60 °C and then heated at 6 °C/min to a final temperature of 200 °C. Injection port and detector temperatures were 250 °C. Peak areas were determined by digital integration after injection of 0.2 microliters (μL) of sample.

Effect of β-Lactoglobulin Level on Benzaldehyde Binding. β-Lactoglobulin (Sigma Chemical Company) and benzaldehyde (Sigma Chemical Company) were combined in 10 mL volumetric flasks to provide duplicate samples of the treatment combinations listed in Table II.

Table II. Treatment Combinations

Treatment	*Benzaldehyde Concentration (mM)*	*% β-Lg (w/v)*
1	4	2.5
2	8	2.5
3	4	5.0
4	8	5.0
5	4	7.5
6	8	7.5

The flasks were sealed with parafilm, heated in a hot water bath at 65 °C for thirty minutes, cooled in ice, and placed on a shaker for 24 hours to equilibrate.

Two mL of each sample was transferred to centrifugal ultrafiltration devices (Amicon MPS-1 Micropartition System) equipped with 10,000 molecular weight cut-off filters (Amicon YM10) to separate free benzaldehyde from protein-bound benzaldehyde (Figure 1).

Free benzaldehyde concentrations were determined by injection of 10 μL ultrafiltrate into an HPLC system equipped with a Constametric 2 G pump (Milton Roy Co.), a Dupont F33588 column (C-18 reverse phase), a Waters Associates Model 441 absorbance detector (λ = 280 nm), and a Hewlett Packard 3390A integrator for determination of peak areas. The flow rate was 1 mL/min. A 0.02 M phosphate buffer was mixed in a 65:35 ratio with HPLC-grade methanol before filtering through a 0.45μ filter (Micron Separations Inc.).

A standard curve was prepared by subjecting protein-free standards to the same heat treatments as the protein-containing samples. The free concentration (mM) of benzaldehyde was determined for each sample by calculation from the standard curve after averaging two injections of each replicate.

Bound benzaldehyde was determined by subtraction [ligand bound (L_b) = ligand total (L_t) - ligand free (L_f)] and percent binding was calculated as L_b/L_t x 100.

Perception of Flavor Compounds in the Presence of Milk Proteins. The samples were prepared and the sensory panelists were trained as described in Hansen and Heinis (*3,4*). All solutions were prepared in a 2.5% sucrose solution. CAS and WPC concentrations were 0%, 0.125%, 0.25%, and 0.5%. The concentrations of flavor compounds were vanillin (78.5 ppm), benzaldehyde (10, 20, and 30 ppm),

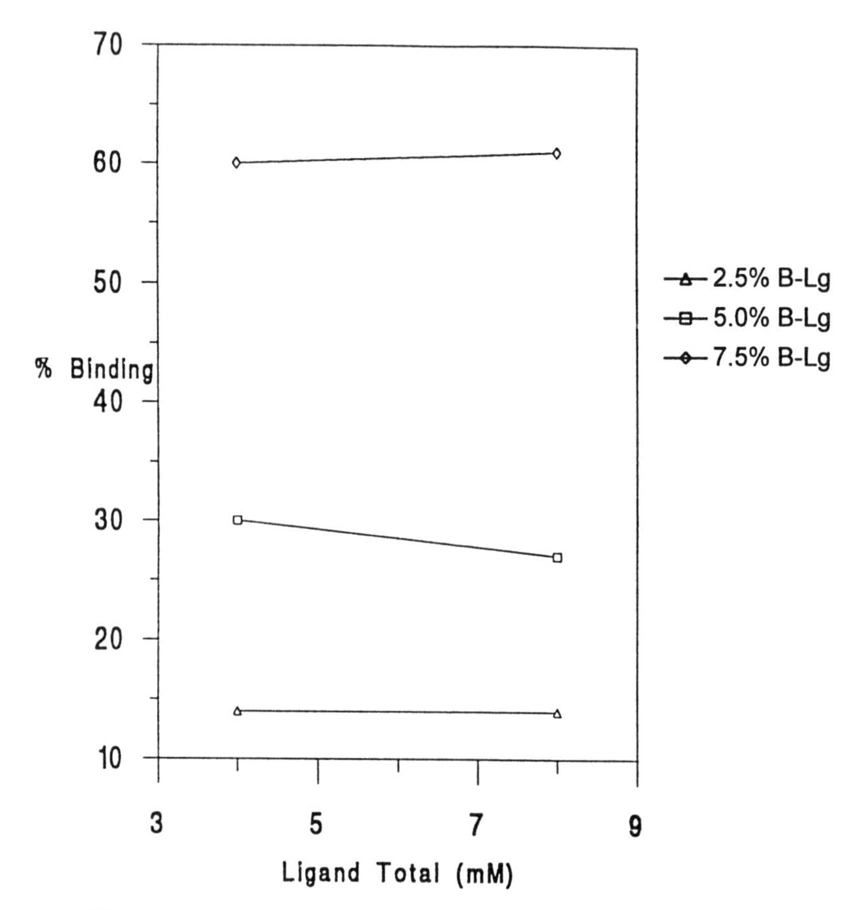

Figure 1. Effect of β-lactoglobulin level on binding of benzaldehyde.

citral (5, 15, and 20 ppm), and *d*-limonene (30, 60, and 90 ppm). The samples were held for 17 hrs at 6 °C to simulate ice cream aging and to allow the protein-flavor compound interactions to equilibrate. The samples were allowed to come to ambient temperature (23 °C).

Samples were presented to the panelists in individual booths illuminated with red light. The panelists were asked to rank the flavor intensity of each sample as compared to a corresponding reference sample. The four references consisted of a 2.5% sucrose solution containing a single flavor compound; vanillin (78.8 ppm), benzaldehyde (17.8 ppm), citral (19.8 ppm), or *d*-limonene (53.0 ppm). The flavor compound levels were chosen to approximate the sensory threshold levels for each. After the samples were rated, the individual flavor and aroma were noted, and samples were retasted and discussed by the panel. Results were analyzed by PROC GLM of SAS statistical software (*11*), and means were compared by the Duncan-Waller procedure.

Results and Discussion

Recovery of Vanillin in Ice Cream Mix Fractions. The amount of vanillin recovered from whey, fat, and casein fractions in samples 1 and 2 is listed in Table III.

Table III. Recovery of Vanillin From Ice Cream Mix Fractions

Sample	*Fraction*	*Weight(g)*	*Vanillin Recovery (mM)*	*% Vanillin Recovered [a] /Total [b]*
1	Whey	36.60	1.57×10^{-3}	48.31[a] /35.19[b]
	Fat	8.79	1.26×10^{-3}	38.77/28.25
	Casein	2.72	4.22×10^{-4}	12.98/9.50
	Total	**48.11**	$\mathbf{3.25 \times 10^{-3}}$	**100.1/73.0**[c]
2	Whey	36.88	2.04×10^{-3}	56.04 [a] /45.33 [b]
	Fat	9.18	1.29×10^{-3}	35.44/28.67
	Casein	2.54	3.13×10^{-4}	8.60/6.96
	Total	**48.60**	$\mathbf{3.64 \times 10^{-3}}$	**100.1/82.6** [c]

[a] mM recovered per fraction divided by total mM recovered.

[b] mM recovered per fraction divided by total mM in the sample.

[c] Total recovery = (mM recovered from all fractions/mM added) x 100.

Total percent recovery (mM recovered/mM added x 100) was 73.0% for sample 1 and 82.6% for sample 2. Of the vanillin recovered, about half was found to be in the whey fraction (48.31% and 56.04%, samples 1 and 2 respectively), while

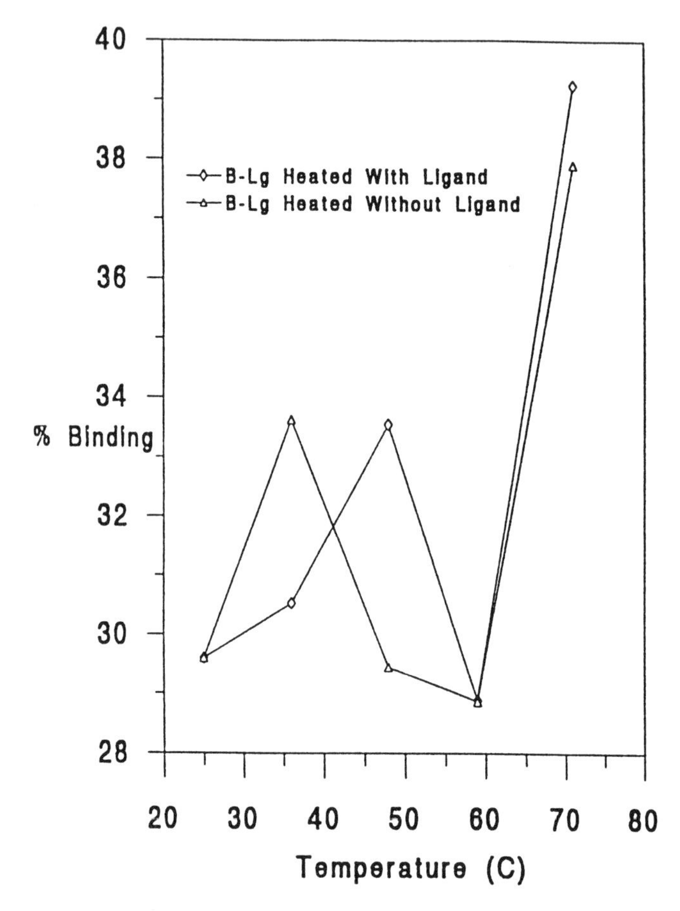

Figure 2. Influence of temperature on binding for β-lactoglobulin heated in the presence or absence of benzaldehyde.

the fat and casein fractions averaged 37.1% and 10.7%, respectively. As a percentage of total composition, the whey fraction is considerably larger than the fat or casein fractions. While most of the flavor remained with the whey, a significant amount was recovered from the fat fraction. The results suggest that a balance between association with whey proteins and partitioning into the lipid phase has occurred.

Effect of β-Lactoglobulin Level on Benzaldehyde Binding. Percent binding versus total ligand concentration is shown in Figure 1 for the three levels of protein (2.5%, 5.0%, and 7.5%) examined. Percent binding appears uniform across total ligand concentration for the two levels investigated (4 mM and 8 mM) and averaged 14.1%, 26.1%, and 60.5% for samples containing 2.5%, 5.0%, and 7.5% protein, respectively.

Influence of Temperature on Binding for β-Lactoglobulin Heated in the Presence or Absence of Benzaldehyde. Percent binding versus temperature for protein solutions heated in the presence or absence of ligand is shown in Figure 2. Percent binding shows a decrease at 59 °C and an increase at 71 °C for samples heated with or without benzaldehyde present during heating.

Effect of Heat on Benzaldehyde Binding to β-Lactoglobulin. All samples pasteurized at 140 °C for 4 seconds (ultra-high temperature (UHT) pasteurization), as well as 12 mM benzaldehyde samples pasteurized at 70 °C for 30 minutes (batch pasteurization), formed a strong gel and could not be analyzed for free benzaldehyde. Ligand-bound versus total plots for both ambient temperature and batch pasteurized samples are shown in Figures 3 and 4, respectively. The slope of the ligand-bound versus ligand total plot for the ambient temperature samples was 0.383, indicating that 38% of the total benzaldehyde was bound by β-lg. Batch pasteurization of samples caused the slope of the ligand bound versus ligand total plot to increase to 0.629, indicating an increase to 63% for amount of benzaldehyde bound. The best-fit line of a double-reciprocal plot of 1/V (moles of protein/moles of bound benzaldehyde) versus $1/L_f$ (1/moles of free benzaldehyde) for the ambient temperature had the following equation:

$$\frac{1}{V} = 0.23464 + \frac{6.0272 \times 10^{-3}}{L_f}$$

indicating a binding constant (K) of 165.9 – M, and 4.2 binding sites (*n*) per monomer of β-lg (Figure 3). Manipulation of data via a double-reciprocal plot for the batch-pasteurized samples yields a negative *y*-intercept and is not valid. Rates of heating were carefully controlled to rule-out temperature induced artifacts. Time versus temperature plots for the 70 °C (batch) and 140 °C (UHT) heat treatments obtained with thermocouples are shown in Figures 5 and 6, respectively.

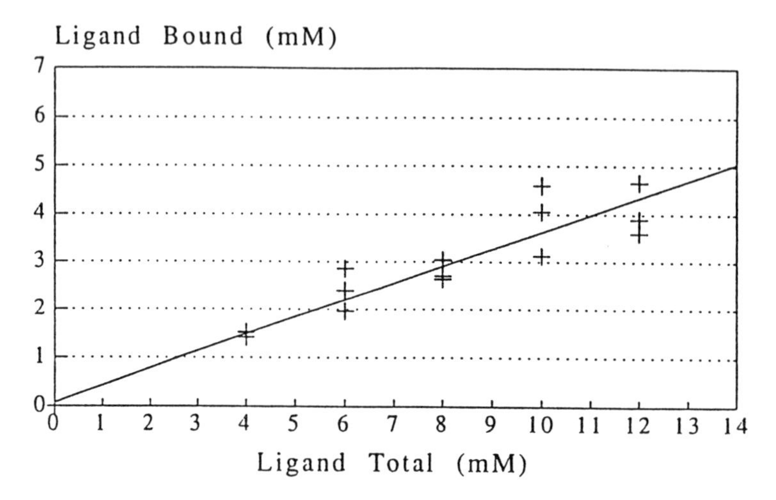

Figure 3. Binding data for ambient temperature samples.

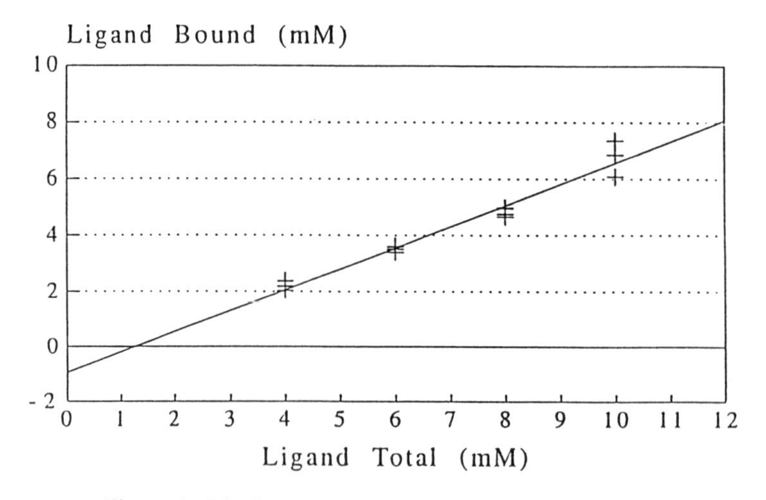

Figure 4. Binding data for batch pasteurization samples.

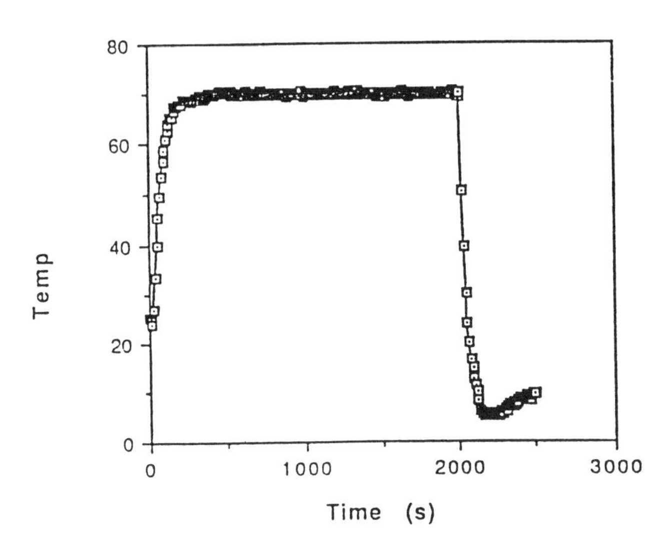

Figure 5. Heating curve for batch-pasteurized samples (70 °C, 30 min).

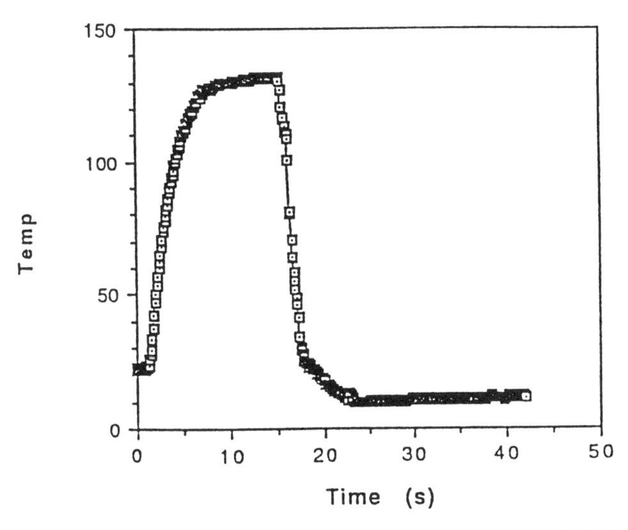

Figure 6. Heating curve for UHT-pasteurized samples (140 °C, 4 s).

Perception of Flavor Compounds in the Presence of Milk Proteins. Figure 7 indicates vanillin flavor intensity relative to the 78.5 ppm vanillin reference standard for different concentrations of CAS and WPC. Figure 8 indicates the intensities of benzaldehyde, citral and *d*-limonene flavor intensity relative to the reference standards (benzaldehyde, 17.8 ppm; citral, 19.8 ppm; or *d*-limonene, 53.0 ppm) at differing concentrations of CAS and WPC. In all cases, a sample with lower perceived flavor intensity than the reference had a flavor score below 0.5

The panelists noted greater flavor intensity of CAS and WPC as the protein concentrations increased.

However, the protein flavor did not overpower the note from the added flavor compound. Vanillin flavor intensity (Figure 7) decreased in the presence of CAS and WPC. Although there was no significant difference in vanillin flavor intensity ($P < 0.05$) with increasing CAS concentration, vanillin flavor declined from 0.32 (moderately less than reference) to 0.15 much less than reference as WPC increased from 0.125 to 0.5%.

Benzaldehyde flavor intensity (Figure 8a) significantly dropped ($P < 0.05$) from 0.45 (slightly less than reference) to 0.25 (much less than reference) as the WPC concentration increased from 0 to 0.5%. There was no significant difference in benzaldehyde flavor concentration as the concentration of CAS increased.

Citral flavor intensity (Figure 8b) showed no significant drop ($P < 0.05$) in intensity as the CAS concentration increased from 0 to 0.5%.

Although citral flavor dropped from 0.41 (slightly less that reference) to 0.33 (less than reference as WPC increased from 0 to 0.5%, this decline was not significant.

d-Limonene flavor intensity (Figure 8c) dropped significantly ($P < 0.05$) in the presence of CAS and WPC. The decline in flavor intensity was most marked for WPC, for which the intensity dropped from 0.41 (slightly less that reference) to 0.27 (much less than reference) as WPC concentration increased from 0 to 0.5%.

Differences in flavor loss between CAS and WPC are expected because of differences in their structure and amino acid composition. The CAS and WPC were subjected to different levels of denaturation due to different processing conditions during manufacturing. Sodium caseinate forms a very loose, open micellar structure with hydrophillic and hydrophobic patches (*12*), which cause it to be highly water soluble and to possess surfactant properties (*13*). Production of CAS causes minimal protein denaturation, even though they are acid precipitated at pH 4.6, washed, resolubilized using sodium hydroxide at pH 6-7, and spray dried (*14*).

Whey protein concentrate consists of a variety of proteins: β-lactoglobulin, α-lactalbumin, immunoglobulins, and bovine serum albumin (*13*). These proteins are heat sensitive and more susceptible to denaturation during processing than casein (*15*). Although BSA can interact with carbonyls (*16*), β-lactoglobulin is most likely to be involved in the flavor compound-WPC interaction because it binds aromatic compounds (*17*).

Vanillin, benzaldehyde and *d*-limonene may interact with the retinol-binding site or with other sites near the surface of the protein (*10*) resulting in decreased concentration of flavor compounds available for perception by the panelist. At

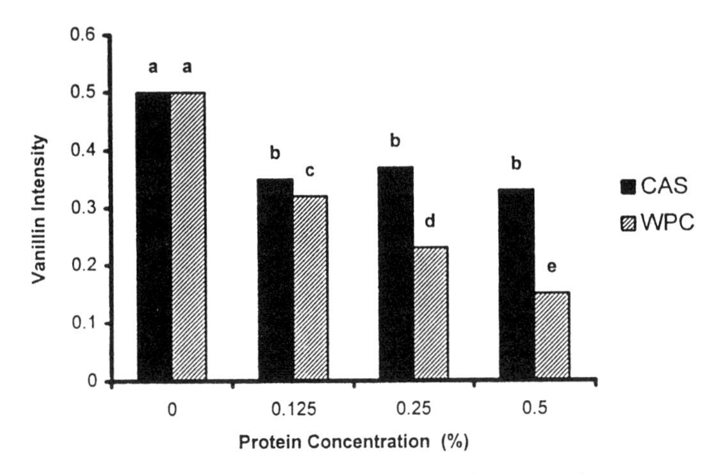

Figure 7. Vanillin flavor intensity relative to reference in the presence of sodium caseinate (CAS) and whey protein concentrate (WPC). Reference vanillin concentration was 3.38 x 10^{-6} mM in a 2.5% sucrose solution. For each protein type, bars with dissimilar letter codes indicate significant differences between means. (Reproduced with permission from ref. 3. Copyright 1991 American Dairy Science Association.)

physiological pH (6.8), retinol binding is most extensive (*12*), and the reactive thiol group of β-lactoglobulin is exposed (*17*), which can react with aldehydes (*18*).

In contrast, a protective effect occurs for citral in citric acid solutions when casein is present (*19*). Citral may form a more open structure in solution so there may be fewer strong interactions within nonpolar binding sites on β-lactoglobulin.

Conclusion

The flavor compounds vanillin, benzaldehyde, citral, and *d*-limonene, showed a tendency to bind with proteins. Binding tended to increase with simultaneous increases in flavor and protein concentration. However, whey proteins exhibited a greater degree of flavor binding than casein. Since the whey proteins are heat sensitive, they unfold with heating and tend to bind a greater amount of flavor compounds. Protein-containing food products that are processed at high temperatures tend to bind more flavor and allow less to be available. As low-fat foods increase in the market, they tend to use higher levels of protein to replace fat and therefore lack flavor. This is especially true if whey proteins are used as fat replacers since they have a greater binding affinity for flavor compounds. The β-lactoglobulin of the whey protein fraction is the most heat sensitive and tends to unfold, thereby allowing benzaldehyde to bind to the protein fraction and reduce the amount available for flavor perception. As the temperature is raised from room temperature to pasteurization temperatures, the amount of binding increases from 38% to 72%.

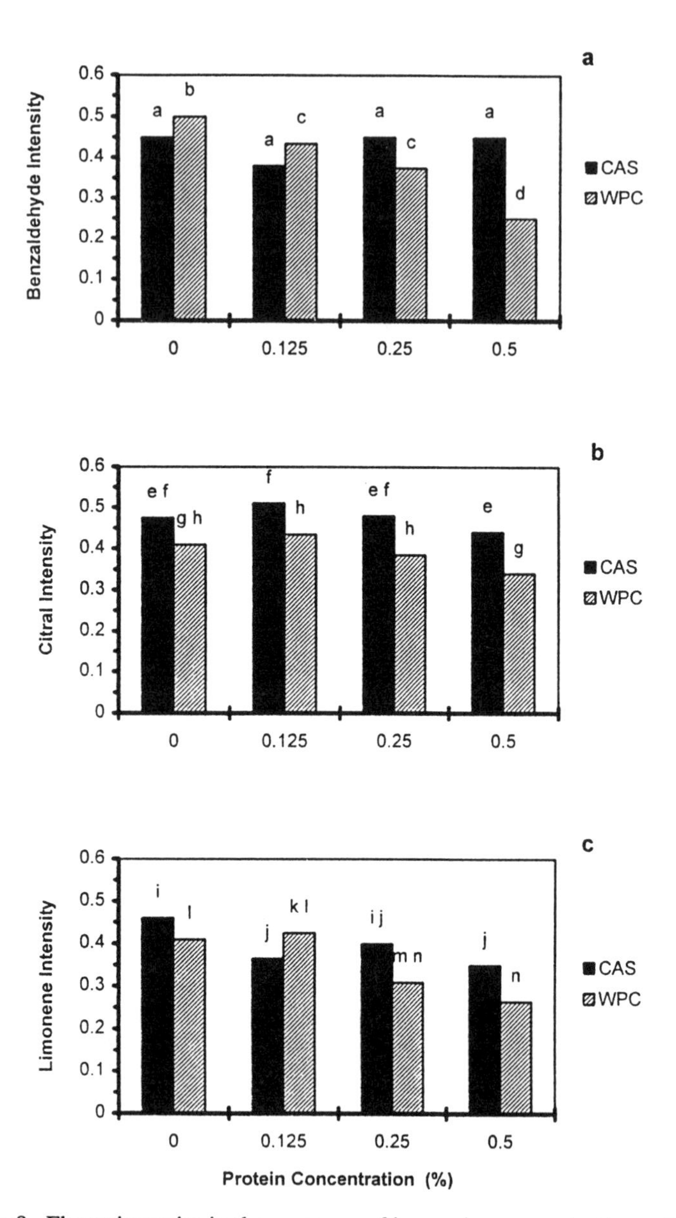

Figure 8. Flavor intensity in the presence of increasing concentrations of sodium caseinate (CAS) and whey protein concentrate (WPC) for benzaldehyde (a), citral (b) and *d*-limonene (c). Reference concentrations are benzaldehyde (0.168 mM), citral (0.130 mM), and *d*-limonene (0.389 mM) in a 2.5% sucrose solution. For each protein type, bars with dissimilar letter codes indicate significant differences between means. (Reproduced with permission from ref. 4. Copyright 1992 American Dairy Science Association.)

References Cited

1. Windholz, M. *The Merck Index*, 9th ed., Merck & Co.: Rahway, NJ, 1976, p. 697.
2. Gildemeister; Hoffman, 1963. *Anterishchen Ole*, IIIc, p. 129.
3. Hansen, A. P.; Heinis, J. J. *J. Dairy Sci.* **1991**, *74*, 2936.
4. Hansen, A. P.; Heinis, J. J. *J. Dairy Sci.* **1992**, *75*, 1211.
5. Duxbury, D. D. *Food Processing*, **1991**, *10*, 43.
6. Defour, E.; Haertle, T. *J. Agric. Food Chem.*, **1990**, *38*, 1691.
7. O'Neill, T.; Kinsella, J. E. *J. Food Sci.* **1988**, *53*, 906.
8. Washnia, A.; Pender, T. W. *Biochem.* **1966**, *5*, 1534.
9. Robillard, K. A.; Washnia, A. *Biochem.*, **1972**, *11*, 3835.
10. Monaco, H. L.; Zanotti, G.; Spadon, P.; Bolognesi, M.; Sawyer, L.; Eiliopoulous, E. E. *J. Mol. Biol.*, **1987**, *197*, 695.
11. *SAS Users Guide: Statistics*, Version 5 ed.; SAS Inst., Inc.: Cary, NC, 1985.
12. Swaisgood, H. In *Milk Proteins*, Barth, C. A.; Schlimme, E., Eds.; Steinkopff Verlag: Darmstadt, Germany, 1989, p. 192.
13. Doxastakis, G. In *Food Emulsifiers, Chemistry, Technology, Functional Properties, and Applications*, Charalambous, G.; Doxastakis, Ed.; Elsevier Science: Amsterdam, The Netherlands; 1989, p. 9.
14. Muller, L. L. In *Developments in Dairy Chemistry*, Fox, I. P. F., Ed.; Applied Science: Barking, England; 1982, p. 325.
15. Morr, C. V.; Foegeding, E. A. *Food Technol.* **1990**, *44*, 100.
16. Damodaran, S.; Kinsella, J. E. *J. Agric. Food Chem.*, **1980**, *29*, 1253.
17. Farrell, H. M., Jr.; Behe, M. J.; Enyeart, J. A. 1987. *J. Dairy Sci.*, **1987**, *70*, 252.
18. Schubert, M. *J. Biol. Chem.*, **1936**, *114*, 351.
19. Friedrich, H.; Gubler, B. A. *Lebensm. Wiss. Technol.*, **1978**, *11*, 215.

Chapter 8

Interaction Between Flavor Components and β-Lactoglobulin

N. Boudaud[1,3] and J.-P. Dumont[2]

Laboratoire d'Étude des Interactions des Molécules Alimentaires, Institut National de la Recherche Agronomique Nantes, [1]B.P. 527, 44026 Nantes Cedex 03, France and [2]B.P. 1627, 44316 Nantes Cedex 03, France

A considerable amount of evidence has been provided in the last thirty years which shows that β-lactoglobulin is able to interact with a variety of ions and small organic molecules. Although this observation is mostly described in terms of flavor binding, it has been generally considered that β-lactoglobulin is a possible carrier for flavor compounds. Structural features shared with proteins in the "odorant-binding" family, and the current interest in predictive models for flavor impact (relating flavor compound availability and intrinsic sensory properties during food consumption) strongly suggested that the macromolecular interaction of β-lactoglobulin with flavor molecules needed to be revisited. In the binding studies described in this chapter, interactions were measured between purified β-lactoglobulin (obtained from homozygous cows possessing variant B) and test compounds selected from different chemical families (diketones, pyrazines, aldehydes, methoxylated benzenes). Experimental data obtained in the static and dynamic modes indicate that, although β-lactoglobulin shows an apparent high affinity for small organic molecules, it may not be very effective in protecting, delivering, or delaying release of most flavor components. It can be concluded that isolated studies of interactions may be misleading if not supported by sensory evidence.

It is generally accepted that sensory-active flavor compounds have to be in the free state to interact with a specific receptor and elicit a sensory response. Simpler systems, as those used in olfactometry, involve partition of the sample between a vapor (sensory-active) and a liquid (sensory-inactive) phase according to the well established laws of physics. Unfortunately, with the exception of beverages, the above model does not readily apply to the bulk of food systems. More often, there is

[3]Current address: Consumer Sciences Department, Institute of Food Research, Earley Gate, Reading RG6 6BZ, United Kingdom

0097–6156/96/0633–0090$15.00/0

no straightforward relationship between the flavor content of the food and the directly available (active) flavor fraction. Investigating the composition of flavor extracts from foods yields valuable information from the qualitative side, but is of little help in providing information on the release of flavor compounds in the time-span of food consumption. Food undergoes tremendous modifications from the first bite to swallowing, caused by mastication and activity of saliva.

Flavor Release – Still Puzzling!

Flavor retention or transient unavailability can be expected to result from mass transfer delayed physically (changes in food microstructure) or chemically (more or less tight binding to food macromolecules). Investigation of physical hindrance caused by food texture is out the scope of this chapter and will not be considered further. However, attention will be focused on aspects dealing with flavor hindrance resulting from binding to food proteins, and the strategies proposed from the early 1970's to link molecular interactions with losses of active flavor compounds.

Interaction vs. Retention – Confusing !

Aside from the well known papers by Arai *et al.* (*1*) and Beyeler and Solms (*2*), the work reported by Franzen and Kinsella (*3*) is a good picture of the repressing effect of macromolecules on the availability of aroma compounds. Gremli (*4*) reporting on the interaction of soy protein with carbonyl compounds came to the conclusion that ketones bind reversibly to the protein while aldehydes bind irreversibly. This suggests that the retention effect must be seen as a complex phenomenon associating loss in flavor potential with delay in the schedule of flavor availability. Later, Mills and Solms (*5*) came to the same conclusion regarding whey proteins. Oddly enough, since the mid-1970's, very little interest appears to have been devoted to flavor retention compared to the bulk of work on molecular interactions. Recently, Ng *et al.* (*6*) reported that sensory perception of vanillin was directly related to the free vanillin concentration in model systems containing fababean proteins. This does not sound unexpected but it is a result that was anticipated for a long time!

At the other extreme of the aroma detection process, proteins were found in the nasal mucosa of dogs and rodents that show surprisingly high affinity for odorants such as 2-isobutyl-3-methoxypyrazine (*7*). It was assumed that proteins could play a role in carrying odorants as albumins do with fatty acids. It was suggested that these carrier proteins could share some characteristic structural features in their sequence and conformation. Strong structural homology existing between β-lactoglobulin and retinol binding protein (*8*) suggested that hydrophobic ligands could be trapped in the β-barrel common to both molecules. The putative role played by β-lactoglobulin in the transportation of vitamin A has inclined to assign the protein to a family alleged to carry small hydrophobic molecules (*9*). β-Lactoglobulin is well represented in bovine milk (2-4 g/L) and in a variety of dairy products. It is known to interact with alkanes (*10, 11*) and volatile carbonyls (*12*) and this makes it potentially either a fantastic functional additive or a troublesome flavor trap.

Evidence presented by O'Neill and Kinsella (*12*) gives support to the assumption that retinol binding to β-lactoglobulin is very specific and cannot be proposed as a model to understand the rather non-specific hydrophobic binding of methyl ketones. This casts some doubt on the validity of propositions, elaborated on the sole grounds of structural analogy with known carrier molecules. Obviously, if the interaction between β-lactoglobulin and flavor compounds has to be taken for granted, the actual influence on flavor perception still needs to be established.

Materials and Methods

Chemicals. Purified β-lactoglobulin (B genetic variant) was obtained using the procedure proposed by Mailliart and Ribadeau-Dumas (*13*). *trans*-Anethole, 2,3,5-trimethylpyrazine (TMP), 2-isobutyl 3-methoxypyrazine, *p*-nitrophenyl phosphate disodium salt, diacetyl, acetophenone and vanillin were commercially obtained from the highest available purity and used without further purification. β-lactoglobulin stock solutions were prepared in trifluoroacetic acid (TFA) (pH 3) and N-[2-hydroxyethyl]piperazine-N'-[2-ethanesulfonic acid] (HEPES) (pH 7.4) buffers. Protein samples were prepared just prior to analysis from stock solutions stored at 4 °C in the dark.

Fluorescence. Emission spectra of the protein were obtained by the static method at 20 °C using a 3-mL thermostated cell. The λ of excitation was 280 nm and usual scan speed was 20 nm/min. Analyzed samples were prepared just before readings by mixing 1.5 mL of protein and ligand stock solutions in the cuvette. Typically, the final protein concentration was 5 μM, while the ligand concentration ranged from 0 to 25 μM. Data from fluorescence were analysed using Stern-Volmer equations (*14*).

$$\frac{F_0}{F} = \frac{1 + K_Q\,[L]}{1 + (1 - f_a)\,K_Q\,[L]}$$

where F_0 and F are the relative fluorescence of protein and the mixture, respectively and f_a is the fraction of initial fluorescence available to the ligand. If f_a is close to 1, a simplified form of the equation can be used.

$$\frac{F_0}{F} = 1 + K_Q\,[L]$$

where [L] is the concentration of the added ligand and K_Q is the quenching constant (M^{-1}) that can be used as a first estimate of K_D, the ligand-macromoledule association constant.

Chromatography. A Kontron HPLC System 600 equipped with a Lichrosorb diol 10 μM column (0.46 x 25 cm) was used. The sample was injected through a 20 μL-loop and monitored by means of an UV detector (Uvikon 730 LC; λ = 280 nm). Injection of a sample containing ligand at the 5-μM level, delivers 0.1 nmole of ligand into the investigated system. The protein concentration ranged from 0 to 200 μM when the external calibration mode was used, and from 0 to 50 μM in the internal calibration mode. External calibration was carried out according to Sun *et al.* (*15*). Samples containing the test flavor molecule are added with different amounts of protein and subsequently injected onto the chromatography column. Deviation from the standard calibration curves obtained for the pure compound gives access to C_f from which C_b can be calculated, since C_t is known:

$$C_b = C_t - C_f$$

where:

C_b = concentration of ligand bound to protein
C_t = concentration of free and bound ligand
C_f = concentration of free ligand

Internal calibration, proposed by Hummel and Dreyer (*16*), was carried out according to Sebille *et al.* (*17*) involving the use of partition chromatography and a cross-linked stationnary phase. Ligand transport by protein results in a negative peak whose area is related to the extent of protein-ligand interaction. It is assumed that information obtained from either mode could be complementary, as equilibration of the presumed complex against pure solvent is achieved in the external mode only.

Results and Discussion

Preliminary experiments showed that the chemical stability of flavor compounds upon storage and over a sufficient time interval to allow measurements was a real concern. *(E)*-Anethole and cinnamaldehyde, which proved to be much too instable as determined by UV absorobance readings over the assay time frame, were rejected from the panel of test components. Acetophenone and vanillin were used for studies in the static mode only, as studies in the dynamic mode involved long equilibration steps prior to chromatographic analysis. Diacetyl eluted too close to the protein to be accurately measured and could not be used in dynamic studies.

Addition of 2,3,5-trimethylpyrazine (TMP), 2-isobutyl 3-methoxypyrazine, or *p*-nitrophenyl phosphate to β-lactoglobulin resulted in quenching of the fluorescence emission of the protein that appeared to increase with the added concentration of pyrazine. Use of the Stern-Volmer equations gave an apparent quenching constant (K_Q) for trimethylpyrazine and isobutyl methoxypyrazine amounting respectively 28000 and 22000 in HEPES buffer. These values are close to that reported for *p*-nitrophenyl phosphate (28800, *14*). When TFA buffer was used, calculated values for trimethylpyrazine and isobutyl methoxypyrazine amounted to 10000 and 12000, respectively. This suggests that the pyrazine ring is important in the formation of the

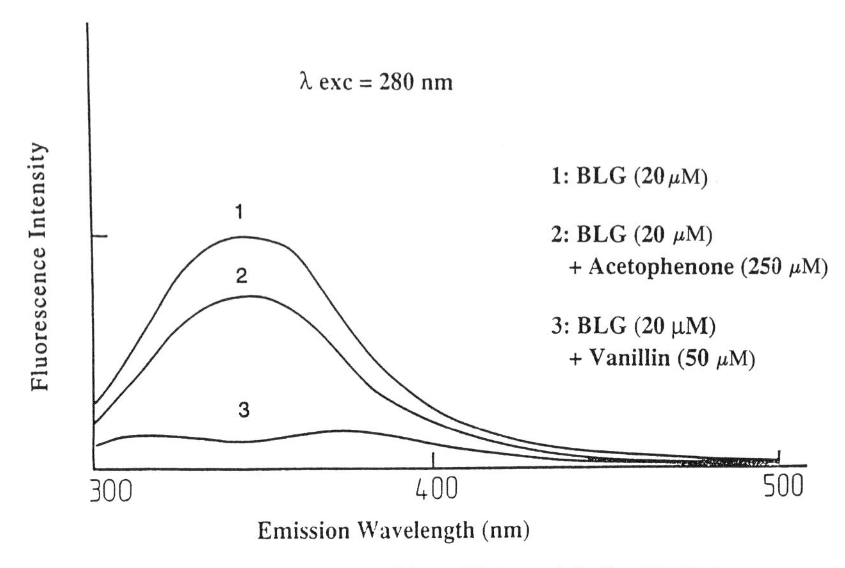

Figure 1. Fluorescence quenching of β-lactoglobulin (BLG) due to interactions with carbonyl compounds (acetophenone, vanillin).

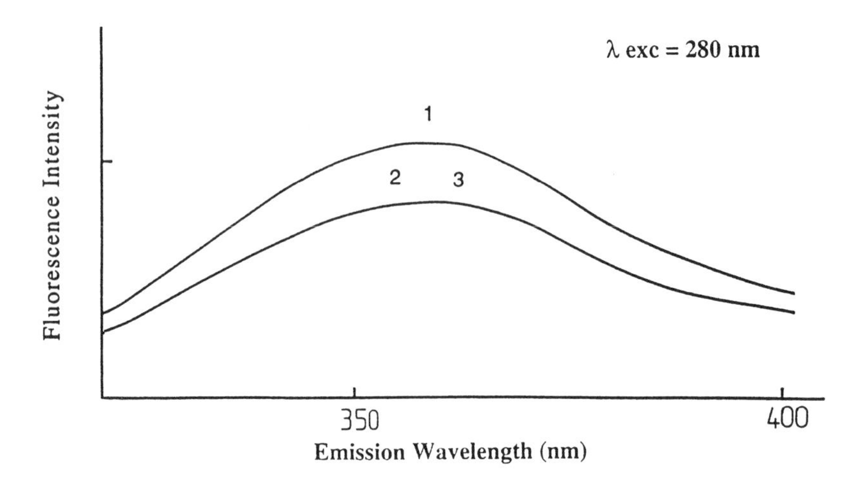

1: N-Acetyl-L-Tryptophane-ethyl ester (10 μM)
2: N-Acetyl-L-Tryptophane-ethyl ester (10 μM) + Benzaldehyde (250 μM)
3: N-Acetyl-L-Tryptophane-ethyl ester (10 μM) + Acetophenone (250 μM)

Figure 2. Fluorescence quenching of N-acetyl-L-tryptophane-ethyl ester with aromatic compounds (benzaldehyde, acetophenone) due to phenyl-phenyl interactions.

complex, while the methoxyl group is not. Regarding interactions with carbonyl compounds, acetophenone and vanillin appeared to be able to quench the fluorescence emission of β-lactoglobulin (Figure 1), but on the contrary, diacetyl showed no effect even at a high molar ratio.

It is more than likely from evidence reported by Mills (*18*) and the value close to unity obtained for f_a, that both tryptophan residues (Trp 19 and Trp 61) of β-lactoglobulin are implicated in an interaction that could be of the phenyl-phenyl type. This was examined using an interaction of model compound N-acetyl-L-tryptophane-ethyl ester with the aromatic compounds, benzaldehyde and acetophenone (Figure 2).

Surprisingly, data obtained in the dynamic mode with the more stable test compounds show no evidence for ligand transport by the protein. It is obvious from Figure 3 that the detector response obtained for trimethylpyrazine is exactly the same whether the injected sample contained β-lactoglobulin or not, leading to the conclusion that no pyrazine has co-eluted with the protein.

In a similar manner, experiments were carried out in the internal calibration mode. Seven concentration levels were chosen for combination of β-lactoglobulin with 2,3,5-trimethylpyrazine as listed in Table I. Injection of the protein in a mobile

Table I. Experimental Parameters for Combination of β-Lactoglobulin with 2,3,5-Trimethylpyrazine

Experiment (Figure 4)	*β-Lactoglobulin Concentration (mM)*	*Trimethylpyrazine Mobile Phase Concentration (mM)*
a	20	6.25
b	50	6.25
c	10	10
d	20	10
e	10	12.5
f	50	12.5
g	10	25

phase containing trimethylpyrazine does not result in any loss of pyrazine at the TMP retention time as shown in Figure 4. It can therefore be concluded that if it had formed, the protein-ligand complex was not stable enough to ensure ligand transport when equilibrated against pure mobile phase. In the light of these results it is doubtful that β-lactoglobulin can compete with the pyrazine-binding protein found in dog nasal mucosa, but the question is still open on the practical aspects of what we call "interactions".

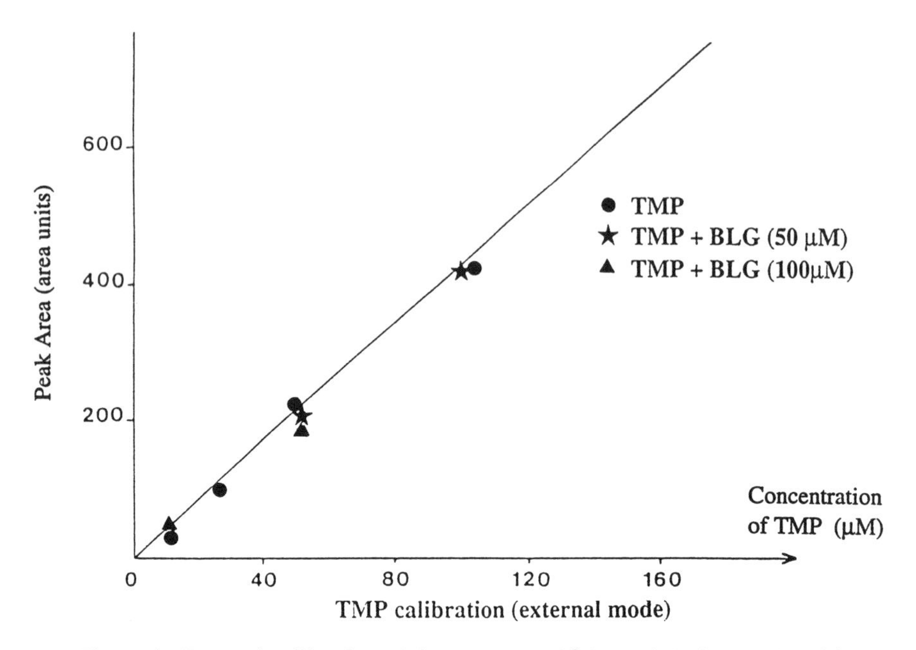

Figure 3. External calibration of the complex of β-lactoglobulin (BLG) with 2,3,5-trimethylpyrazine (TMP).

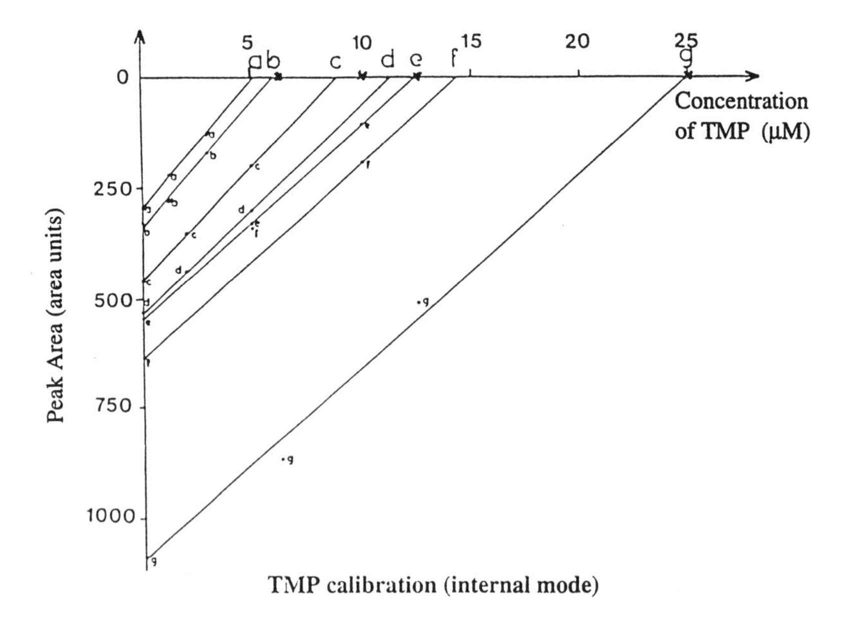

Figure 4. Internal calibration of the complex of β-lactoglobulin (BLG) with 2,3,5-trimethylpyrazine (TMP). Experimental parameters provided in Table I.

Conclusion

Analysis of protein fluorescence provides a quick and elegant means to trace complex formation with flavor components in a static situation, and can help to identify molecular interactions. A major drawback of the technique appears when flavor availability during food consumption, rather than flavor retention in a model system, is to be measured. Chromatographic measurements, carried out in a changing environment, seem to be more receptive to investigate ligand exchanges in ingested food products. Nevertheless, prediction of flavor loss remains difficult for low-affinity complexes between flavor compounds and the macromolecule. β-Lactoglobulin, while able to unambiguously interact with 2,3,5-trimethylpyrazine, 2-isobutyl 3-methoxypyrazine and *p*-nitrophenyl phosphate through hydrophobic phenyl bonds, appears to be unable to carry these ligands when it is equilibrated against pure aqueous solvents. This suggests that, although it is structurally related to retinol-binding and odorant-binding proteins, β-lactoglobulin may not necessary be functionally related to them.

Literature Cited

1. Arai, S.; Noguchi, M.; Yamashita, M.; Kato, H.; Fujimaki, M. *Agric. Biol. Chem.* **1970**, *34*, 1569.
2. Beyeler, M.; Solms, J. *Lebensm. Wiss. Technol.* **1974**, *7*, 217.
3. Franzen, K. L.; Kinsella, J. E. *J. Agric. Food Chem.* **1974**, *22*, 675.
4. Gremli, H.A. *J. Am. Oil Chem. Soc.* **1974,** *5*, 95A.
5. Mills, O. E.; Solms, J. *Lebensm. Wiss. und Technol.* **1984,** *17*, 331.
6. Ng, P. K. W.; Hoehn, E.; Bushuk, W. *J. Food Sci.* **1989**, *54*, 324.
7. Pevsner, J.; Sklar, P. B.; Snyder, S. H. *Proc. Natl. Acad. Sci. USA*. **1986**, *83*, 4942.
8. Godovac-Zimmerman, J. *T.I.B.S.* **1988,** *13*, 64.
9. Pervaiz, S.; Brew, K. *Science* **1985**, *228*, 335.
10. Mohammadzadeh, A.; Feeney, R. E.; Smith, L. M. *Biochem. Biophys. Acta.* **1969**, *194*, 246.
11. Mohammadzadeh, A.; Feeney, R. E.; Smith, L. M. *Biochem. Biophys. Acta.* **1969**, *194*, 256.
12. O'Neill, T. E.; Kinsella, J. E. *J. Agric. Food Chem.* **1987**, *35*, 770.
13. Mailliart, P.; Ribadeau-Dumas, B. *J. Food Sci..* **1988**, *53*, 743.
14. Farrell, H. M.; Behe, M. J.; Enyeart, J. A. *J. Dairy Sci..* **1987**, *70*, 252.
15. Sun, S. F.; Kuo, S. W.; Nash, R. A. *J. Chromatogr.* **1984**, *288*, 377.
16. Hummel, J. P.; Dreyer, W. J. *Biochim. Biophys. Acta.* **1962**, *63*, 530.
17. Sebille, B.; Thuaud, N.; Tillement, J. P. *J. Chromatogr.* **1979**, *180*, 103.
18. Mills, O. E. *Biochim. Biophys. Acta.* **1976**, *434*, 324.

Chapter 9

The Effect of Gelling Agent Type and Concentration on Flavor Release in Model Systems

James Carr[1], David Baloga[2,4], Jean-Xavier Guinard[3], Louise Lawter[2], Cecile Marty[3], and Cordelia Squire[1]

[1]Food Ingredients and Additives Group, Systems Bio-Industries, Inc., 620 Progress Avenue, Waukesha, WI 53187
[2]Flavor and Fruit Group, Systems Bio-Industries, Inc., 2607 Interplex Drive, Trevose, PA 19053
[3]Department of Nutrition, Pennsylvania State University, University Park, PA 16802

Differences in flavor perception are sometimes observed when product stabilization systems are altered. To verify this hypothesis, the flavor release of three fruit flavored gels at three gel strengths was examined using analytical and sensory methodologies. Three fruit flavors of different chemical classes, i.e. cherry (aromatic aldehyde), citrus (monoterpene), and grape (ethyl ester), and gelling agents with different chemical structures (gelatin, starch, carrageenan) were selected to study the effect of chemical interaction on flavor release. Flavor perception of gels composed of three flavor/stabilizer combinations was evaluated at low, medium and firm gel strengths to examine the role rheology played in flavor release. A trained sensory panel was used to measure flavor intensity difference among samples. The model gels were analyzed via quantitative dynamic headspace analysis using a sample preparation technique that simulated flavor release in the mouth. Sensory and analytical data were correlated. A decrease in flavor perception was observed with increase in gel strength. Perceived flavor intensity was also found to be dependent on the type of gelling agent.

Flavor perception in foods is often strongly affected by a variety of non-flavor food components. Hydrocolloids present in food products may directly or indirectly influence the rate and intensity of flavor release. This may be due to a physical entrapment of flavor molecules within the food matrix or may be caused by a specific or non-specific binding of flavor molecules. A detailed review of the possible interactions between classes of volatile flavor compounds and different food system components such as lipids, carbohydrates, proteins, purines and phenols has been previously reported (*1,2*).

[4]Current address: Quest International, 5115 Sedge Boulevard, Hoffman Estates, IL 60192

0097–6156/96/0633–0098$15.00/0

Trained sensory panels have been used to study the effects of gelatin concentration on flavor perception (*3*). The authors found that gelatin concentration modified the flavor perception. They concluded that while specific physical interactions of gelatin and flavor compounds may have occurred, it was more likely that the observed results were due to differences in the mechanical strengths of the gels tested.

In an examination of a number of different gelling agents, it was found that hardness negatively correlated with overall fruit flavor intensity (*4*). Interestingly, gelatin showed exceptional overall flavor considering its hardness value. Other samples showed lower overall flavor intensity than would have been predicted by hardness value. These results could therefore suggest that other non-texture factors may influence the release of flavor in food systems.

Although different types of gelling agents exhibit different rheological properties, an effort can be made to prepare approximately equivalent gels in terms of single analytical attributes. In the work presented here, a variety of different gelling agents were used to prepare gels with soft, medium and firm gel strengths (as measured by gel hardness). These gels contained one of three different common flavor compounds of different chemical classes (aromatic aldehyde, monoterpene and ethyl ester). Dynamic headspace chromatography and sensory evaluation were used to examine the nature of the interaction between flavor and texture.

Experimental

Flavor compounds. Benzaldehye, 99%, ethyl butyrate, 99%, and *d*-limonene, 97.5% were obtained from Aldrich Chemical Company, Inc. (Milwaukee, WI) and chromatographed (GC-FID) for respective purities. Each flavor compound was quantitatively diluted in 95% (v/v) ethyl alcohol such that a 1mL-aliquot delivered 10 ppm, 60 ppm, and 180 ppm of benzaldehyde, ethyl butyrate or *d*-limonene, respectively, to the gel matrix. The flavor concentrations were established by a consensus of four experienced flavorists and organoleptically characterized as generic cherry (benzaldehyde), grape (ethyl butyrate) and citrus (*d*-limonene) flavor.

Gel model system. Gelatin (250 Bloom type A 40 mesh) and carrageenan (unstandardized commercial iota type) gelling agents were obtained from Sanofi Bio-Industries, Inc. (Waukesha, WI). Corn starch (Miragel® 463) was obtained from National Starch, Inc. (Bridgewater, NJ) . Gelling agents were used on an as-is basis.

The composition of the model dessert gel is shown in Table I. Gel samples were prepared by first preblending sucrose (75.0 g), tripotassium citrate (1.0 g), citric acid (2.0 g) and the appropriate amount of gelling agent. This dry blend was then added to a Waring Blendor containing deionized water of sufficient quantity to yield a final weight of 500.0 g. The gel solution was mixed for 1 min at high speed. The flavor solution was dispersed using a 1.0-mL pipette while continuously mixing for an additional 1 min. The samples were quantitatively transferred to 800-mL glass jars and hermetically sealed. The samples were heated under slight agitation for 15 min at 100 °C, then transferred to a 25 °C bath and statically cooled for 15 min. Samples were transferred to 5 °C storage and held for 16 h prior to testing.

Table I. Dessert Gel Model Composition

Component	*Amount (%)*
Deionized water	78.38–83.20
Sucrose	15.00
Tripotassium citrate	0.20
Citric acid	0.40
Gelling Agent	1.20–6.00
Flavor compound	0.001–0.018

The gel model systems examined in this study are described in Table I. Concentrations for the gelling agents were determined by choosing the level at which each gave approximately equivalent gel hardness, as defined in Texture Analysis section, at three overall levels (soft = 0-50 g, medium = 50-100 g, firm = 100-150 g). Flavor compounds were used at the characterizing concentrations as noted above.

Texture Analysis. Gels were prepared for texture analysis using the gel model system procedure described above. Prior to cooling, samples were transferred to 100-mL Pyrex crystallizing dishes and covered with plastic lids. The samples were stored at 5 °C for 16 h prior to analyses.

Texture analysis was conducted using a TA.XT2 Texture Analyzer (Texture Technologies, Inc., Scarsdale, NY) interfaced to a personal computer. Testing involved the generation of force-distance curves using a 12.7-mm plunger at 0.5 mm/second compression speed to a distance of 20 mm. A variety of different parameters were examined with force at rupture being reported as gel hardness.

Sensory Analysis. Bench testing of the gels before the study showed that the time from sample intake to swallowing typically varied between 4 and 15 seconds depending on the size and on the firmness of the gels. These observations were taken into account when designing the sensory evaluation. A trained panel consisting of twelve judges rated flavor intensity of the gels in duplicate on a 145 mm visual analog scale anchored with terms "low intensity" and "high intensity". Judges were trained to rate flavor intensity with water solutions of three flavors at low, medium and high concentrations. In three subsequent sessions conducted on three consecutive days at the same time of day, judges rated the flavor intensity of the gels. One flavor was rated per session (e.g. *d*-limonene, benzaldehyde or ethyl butyrate). In each session, the judges evaluated the 9 samples (3 gelling agents x 3 gel strengths) as 3 sets of 3 samples presented in a randomized order. Judges were instructed to chew the gels served as 25-g samples in 2-oz plastic cups for about 5 seconds and to rate the intensity of the their flavor. Judges rinsed with water, ate a salt-free cracker and waited for 3 min between sets. This evaluation was conducted in individual booths under incandescent light. Analysis of variance (ANOVA) was performed using SPSS software with judges, replications, gelling agent (carrageenan, gelatin, starch), concentrations (soft, medium, firm), and the two-way interaction terms as sources of variation. Fisher's LSD at the 0.05 significance level was used to compare mean flavor intensity data.

Dynamic Headspace Sampling – Gas Chromatography/Mass Spectrometry. Samples were ambiently (ca. 22 °C) equilibrated for 15 min just prior to analysis. A 100-g gel sample was placed in a 250-mL beaker and stirred at 300 rpm for 60 s using an IKA stirrer (model RW20 DZM, IKA-Werk, Inc.) equipped with a 4-blade 50-mm propeller (model R1342, IKA-Werk, Inc.). A 20-g sample of the comminuted gel was placed in a 25 mL sparge vial (Tekmar, Cincinnati, OH). The samples were preheated and prepurged with helium at 60 cc/min for 1 min, purged for 30 seconds at 60 cc/min at 37 °C onto a Tenax trap and then desorbed at 180 °C for 3 min. Gas chromatography was conducted using a Hewlett-Packard (Avondale, PA) 5890 Series II GC directly interfaced via jet separator to Hewlett-Packard (Avondale, PA) 5971 mass spectrometer. The GC column used was a 60 m x 0.53 mm i.d. x 3.0 μm film thickness DB-624 (J&W Scientific, Folsom, CA). Inlet and transfer line temperatures were 180 °C and 250 °C, respectively. Helium was used as the carrier gas at an average linear velocity (μ) of 40 cm/s (10 °C). Flavor volatiles were separated by cryogenically cooling the oven to an initial temperature of 10 °C (3 min hold). The oven was ramped at 10 °C/min to 180 °C then ramped at 70 °C/min to 250 °C (4 min hold). The mass spectrometer was operated in the electron impact mode at 70 eV scanning form 29 to 450 *m/z* in a 1.3-s cycle.

Headspace concentration of the flavor compounds was calculated using a 3-point external standard calibration. Flavor compounds were diluted over the experimental range in deionized water and analyzed using the above procedure. The results were expressed in ppm.

Results and Discussion

A closed model system for the evaluation of texture-flavor interactions was developed. Table II summarizes the experimental variables and results from texture analyses for all 27 samples. Increased gel hardness reduced flavor intensity perception and the quantity of flavor compound released into the sample headspace for all three gelling agents examined (gelatin, carrageenan and starch). Figure 1 describes sensory flavor intensity versus gel hardness for each of the experimental conditions (gelling agent type, gelling agent usage level and flavor compound type). Different gelling agents affected this flavor release to differing extents.

Effect of Gelling Agent on Flavor Intensity Perception. Table III describes Mean Flavor Intensity ratings for each gelling agent and gel strength examined. Gelatin and carrageenan showed the highest perceived flavor intensities for the soft and medium hardness gels for all three flavor compounds, while starch showed the lowest overall flavor intensity for all three flavor compounds for the soft and medium hardness gels. Hard gels composed of carrageenan showed the highest

Table II. Description of Experimental Variables

Sample #	*Gel Agent Type*	*Gel Agent Use Level (%)*	*Flavor Compound*	*Flavor Use Level (ppm)*	*Gel Hardness*
1	Gelatin	1.2	Benzaldehyde	60	38.3
2	Gelatin	1.2	Ethyl butyrate	10	38.3
3	Gelatin	1.2	*d*-Limonene	180	38.3
4	Gelatin	1.6	Benzaldehyde	60	71.5
5	Gelatin	1.6	Ethyl butyrate	10	71.5
6	Gelatin	1.6	*d*-Limonene	180	71.5
7	Gelatin	2.0	Benzaldehyde	60	145.2
8	Gelatin	2.0	Ethyl butyrate	10	145.2
9	Gelatin	2.0	*d*-Limonene	180	145.2
10	Carrageenan	1.2	Benzaldehyde	60	32.9
11	Carrageenan	1.2	Ethyl butyrate	10	32.9
12	Carrageenan	1.2	*d*-Limonene	180	32.9
13	Carrageenan	1.4	Benzaldehyde	60	57.2
14	Carrageenan	1.4	Ethyl butyrate	10	57.2
15	Carrageenan	1.4	*d*-Limonene	180	57.2
16	Carrageenan	1.8	Benzaldehyde	60	142.9
17	Carrageenan	1.8	Ethyl butyrate	10	142.9
18	Carrageenan	1.8	*d*-Limonene	180	142.9
19	Starch	4.0	Benzaldehyde	60	42.6
20	Starch	4.0	Ethyl butyrate	10	42.6
21	Starch	4.0	*d*-Limonene	180	42.6
22	Starch	5.0	Benzaldehyde	60	76.0
23	Starch	5.0	Ethyl butyrate	10	76.0
24	Starch	5.0	*d*-Limonene	180	76.0
25	Starch	6.0	Benzaldehyde	60	104.4
26	Starch	6.0	Ethyl butyrate	10	104.4
27	Starch	6.0	*d*-Limonene	180	104.4

Table III. Mean Flavor Intensity Ratings on a 145-mm Scale for Different Gelling Agents and Gel Strengths (N=12)

Flavor Compound	*Medium*	*Gel Strength*		
		Soft	*Medium*	*Hard*
Benzaldehyde	Starch	88.3	55.8	48.9
	Gelatin	108.9	89.4	42.4
	Carrageenan	103.8	82.3	51.5
d-Limonene	Starch	73.7	38.2	45.3
	Gelatin	116.9	84.5	71.5
	Carrageenan	92.5	87.5	45.0
Ethyl butyrate	Starch	68.8	44.3	43.8
	Gelatin	92.3	90.5	43.9
	Carrageenan	106.1	88.9	63.9

flavor intensity for gels flavored with benzaldehyde and ethyl butyrate. Gelatin hard gels showed the highest flavor intensity for *d*-limonene (Figures 2-4). Gelling agent type and gel concentration were a significant source of variation ($p < 0.01$) for all three flavors. Replication was not a significant source of variation.

The differences in perceived flavor intensity between starch and either gelatin or carrageenan was less pronounced for benzaldehyde. Gels flavored with both ethyl butyrate and *d*-limonene showed a similar trend with differences in perceived flavor intensity being greater for gels composed of carrageenan and gelatin compared to starch gels. No strong differences were observed among flavor compounds for a given texture agent.

Effect of Gelling Agent on Flavor Compound Headspace Concentration. Table IV lists the relative headspace concentrations of each flavor compound for each gelling agent and gel strength examined. Quantitative dynamic headspace analysis results correlated with trained sensory panel data for flavor intensity. Increasing gel hardness reduced the amount of flavor compound present above the comminuted samples (Figures 5-7). As was the case for sensory panel flavor intensity measurements, gelatin provided the highest flavor release, followed by carrageenan. With the exception of ethyl butyrate flavored hard gels, starch based gels exhibited the lowest flavor release into the headspace as gel hardness was increased.

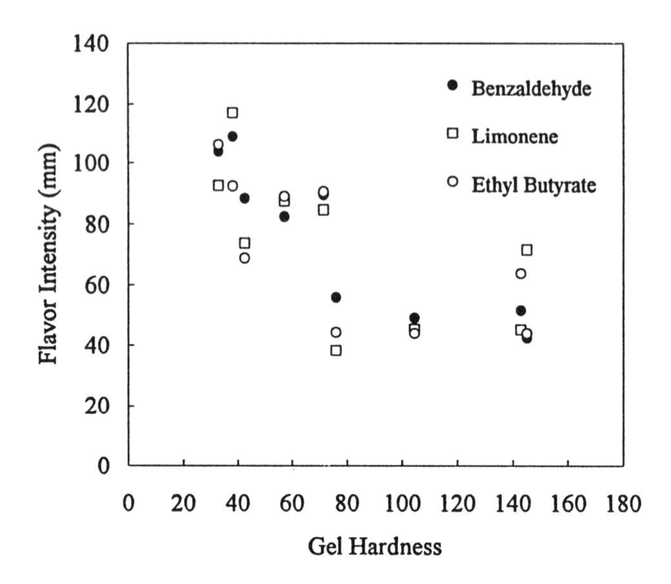

Figure 1. Effect of gel hardness on flavor intensity.

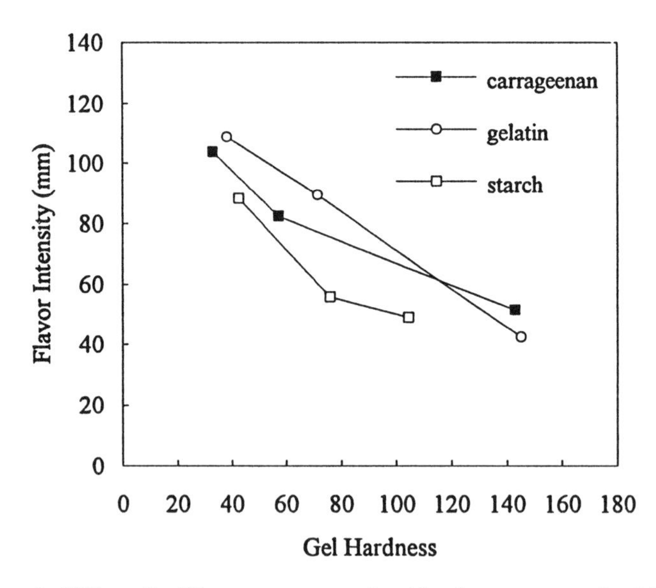

Figure 2. Effect of gelling agent type and gel hardness on perceived flavor intensity of benzaldehyde.

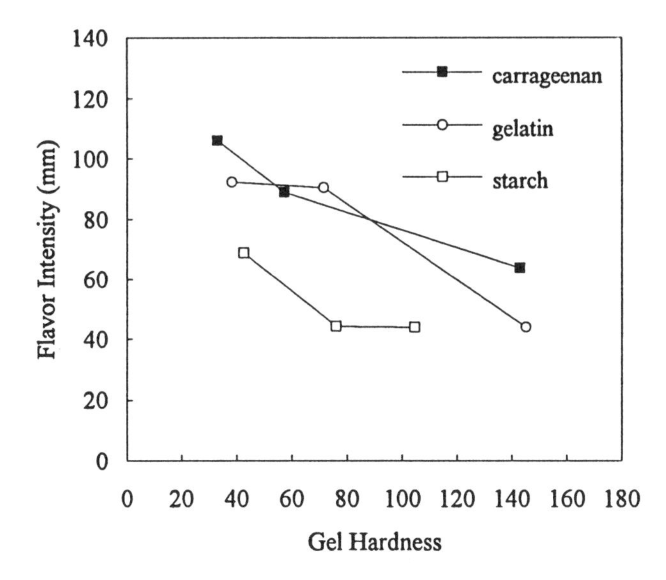

Figure 3. Effect of gelling agent type and gel hardness on perceived flavor intensity of ethyl butyrate.

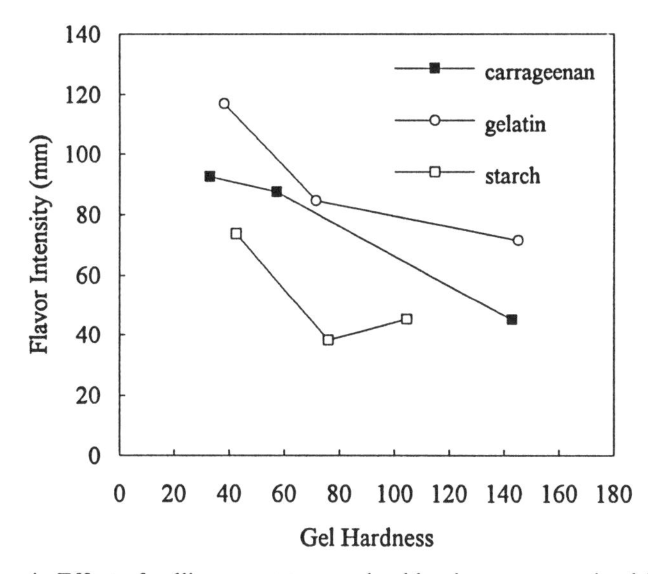

Figure 4. Effect of gelling agent type and gel hardness on perceived flavor intensity of limonene.

Table IV. Relative Headspace Concentration for Different Gelling Agents and Gel Strengths

Flavor Compound	*Medium*	*Relative Headspace Concentration (ppm)*		
		Soft	*Medium*	*Hard*
Benzaldehyde	Starch	9.2	5.6	0.8
	Gelatin	25.2	14.5	1.1
	Carrageenan	23.6	8.2	3.6
d-Limonene	Starch	8.3	4.3	1.2
	Gelatin	29.3	20.6	2.7
	Carrageenan	12.8	6.9	2.2
Ethyl butyrate	Starch	6.8	5.6	3.8
	Gelatin	9.0	6.8	3.9
	Carrageenan	5.3	4.5	4.2

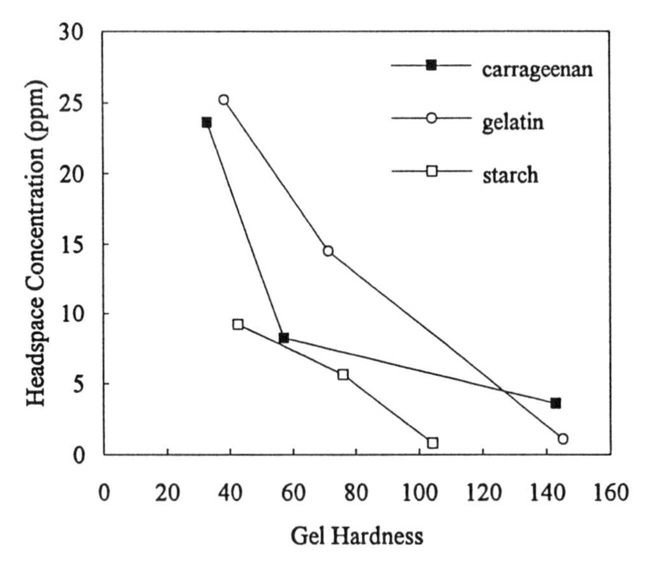

Figure 5. Effect of gelling agent type and gel hardness on headspace concentration of benzaldehyde.

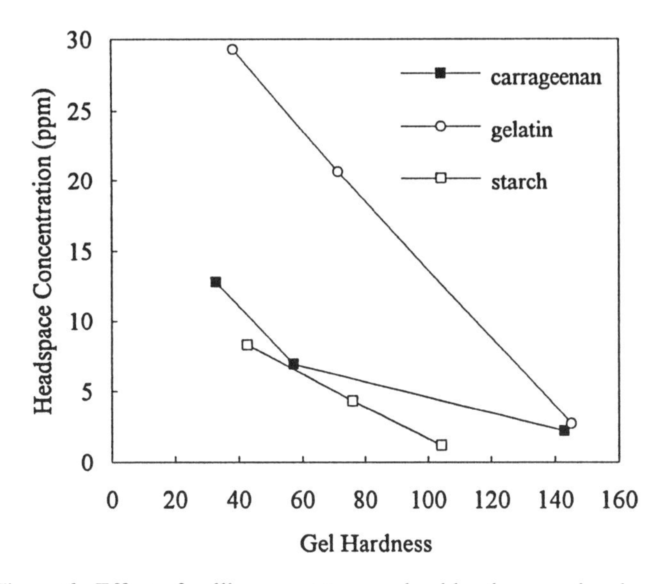

Figure 6. Effect of gelling agent type and gel hardness on headspace concentration of limonene.

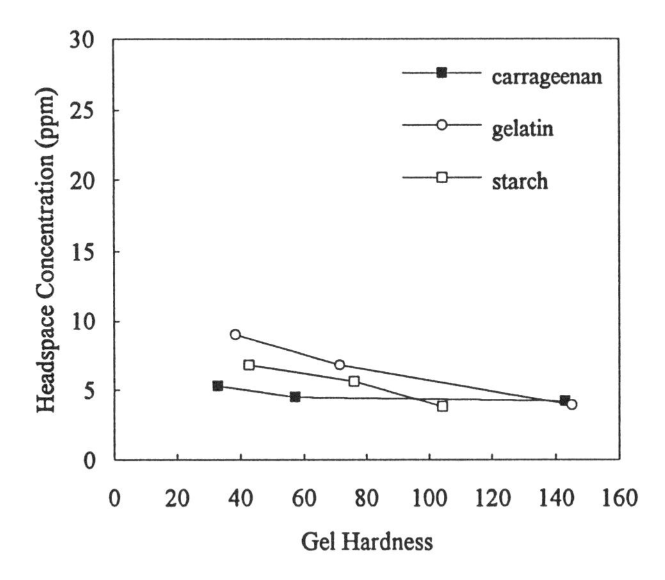

Figure 7. Effect of gelling agent type and gel hardness on headspace concentration of ethyl butyrate.

Conclusions

High flavor impact is important in developing acceptable and cost-effective flavor profiles in processed food products. The texturing agent as well as the overall rheological characteristics of the system must be considered, however, if this flavor profile goal is to be met. As this study suggests, an increase in gelling agent will require an increase in flavor concentration to deliver the same flavor impact. In addition, the specific gelling agent may influence the optimum flavor use level.

The change in consistency for most foods during consumption involves the comminution of the food in the oral cavity, softening as a result of a temperature rise and a moisture uptake due to the presence of saliva (*5*). The extent to which the consistency of the food decays during this process can affect not only the textural properties of the food in the mouth, but also the flavor perception for the food.

Textural attributes at body temperature, while in the mouth, may account for the high observed flavor intensity properties of gelatin. Gelatin softens and exhibits a melting behavior between 30-35 °C and this could promote a release of flavor molecules more easily from the food matrix under these conditions.

Carrageenans are often used in dessert gel systems to provide a texture approaching that of gelatin. While the texture can approximate the elastic characteristics of the target, the melt profile will be different since carrageenan gels melt at higher temperature. This may account for the observed lower flavor intensity for a number of the model gels prepared in this study.

Starch is often used at higher concentrations in food products than other common hydrocolloids (i.e., several percent). At these levels where the desired functional properties are achieved, some evidence of reduced flavor intensity and release are suggested.

While no overall differences in the flavor intensity among the three flavor compounds were apparent, however considering the vast number of flavor compounds available, more flavor compounds with a wider range of volatility and functionality need to be investigated. Variation in flavor release of different flavor compounds with the same gelling agent would explain flavor imbalances in multi-component flavored gels.

Literature Cited

1. Solms, J; Osman-Ismail, F.; Beyeler, M. *J. Can. Inst. Food Sci. Technol.* **1973**, *6*, A10-A16.
2. Solms, J.; Guggenbuehl, B. In *Flavour Science and Technology*; Bessiere, Y., Thomas, A.F.; Eds.; John Wiley & Sons, Inc.: Chichester, England, 1990; pp 319-335.
3. Jaime, I; Mela, D. J.; Bratchell, N. *J. Sensory Studies* **1993**, *8* (3), 177-188.
4. Clark, R. In *Frontiers in Carbohydrate Research* ; Chandrasekaran, R., Ed.; Elsevier Applied Science: New York, NY, 1992, Vol. 2, pp 85-89.
5. de Bruijne, D. W.; Hendrickx, H. A. C. M.; Alderliesten, L.; de Looff, J. In *Food Colloids and Polymers: Stability and Mechanical Properties*; Dickinson, E.; Walstra, P.; Royal Society of Chemistry: Cambridge, 1993; pp 204-213.

Chapter 10

Binding of Volatiles to Starch

M. Y. M. Hau, D. A. Gray, and A. J. Taylor

Department of Applied Biochemistry and Food Science, University of Nottingham, Sutton Bonington Campus, Loughborough LE12 5RD, United Kingdom

The binding of aroma volatiles to starch has been studied in a closed system. Volatiles were introduced into the headspace and binding was measured by assaying the headspace concentration using gas chromatography. Factors affecting the reproducibility of the analysis are discussed and the analysis is compared with other methods for measuring binding. The described system uses volatiles at concentrations found in food systems and allows the study of parameters that affect binding. Examples of these parameters are the geometry of the sample (e.g., surface area), the physical state (glassy and rubbery) and the volatile concentration. The system has a wide range of uses in measuring binding (or release) of volatiles to (or from) different types of food materials under conditions that are similar to those found in real foods. Further analysis of the data may allow the mechanisms of volatile binding to be elucidated.

The various interactions of flavor molecules with materials such as starch are responsible for affecting the flavor characteristics of food products (*1*). Different flavors bind to food materials to various extents with distinct physicochemical mechanisms. Binding of flavors tends to suppress their impact and/or perception and frequently results in an imbalance in the flavor profile (*2,3*). The flavor impact as perceived by the consumer is important in determining the acceptability of foods, hence the phenomenon of flavor binding and release is extremely significant. Equally, understanding these interactions would assist in controlling flavor loss during storage as well as flavor release during consumption.

The mobility of flavor molecules in foods is known to affect the flavor characteristics. It is desirable to minimize such movement during processing and subsequent storage to avoid flavor loss. A glass transition in a food material is thought to exert a profound effect on the ability of small molecules such as flavors

0097–6156/96/0633–0109$15.00/0

to penetrate the food matrix. The glass transition is the onset of segmental mobility in the amorphous or semi-crystalline polymer, where the molecules are no longer considered immobile but have sufficient energy to slide past one another (*4*). The polymer becomes viscous, rubbery and flexible. The glass transition temperature increases with increased degree of crystallinity (*5*) and is known to decrease with increasing concentrations of plasticizer such as water (*6*) and other low molecular weight substances (*7*). Solid carbohydrates can exist in either a stable crystalline or a metastable amorphous state, where the metastable amorphous form will transform to the crystalline under suitable conditions. This transformation will have a significant effect on the properties and quality of food products. Thus, knowledge regarding the conditions leading to the production of desired structures and to the retention of these structures is important. Starch is an important food component which exhibits a glass transition and so can exist in either a glassy or rubbery form. The mobility of flavor molecules through the starch matrix in the rubbery state will be greater due to the increased free volume caused by the substantially greater molecular motion.

Measuring the movement of molecules in food systems is complicated by the fact that these systems can vary widely in composition and texture. They can vary from homogeneous aqueous solutions to heterogeneous systems composed of water and lipid phases or partially dissolved or undissolved carbohydrates, proteins etc. Therefore a variety of methods must be applied in characterizing the binding process. In the early 1970's, dialysis equilibrium methods (*1*) were often used for binding studies. Currently, headspace methods are favored which measure the changes in headspace concentration caused by the presence of a food ingredient.

Carbohydrate / flavor interactions have been studied by Maier (*8,9*) who has used numerous methods including headspace methods to determine the sorption of flavors by solid food components. Subsequent experiments (*10-12*) studied the adsorption of volatile aroma compounds to, and their desorption from, foods in desiccators containing a mixture of solutes and salts to provide a series of vapor pressures over the food from which kinetic sorption curves were obtained by weighing the solid after certain time intervals. Spectroscopic methods (*13*) in which thin films of food materials were subjected to IR spectroscopy before and after sorption of volatile substances were also used. Numerous studies have been performed on the interaction of low boiling aroma components with biopolymers of low moisture content (*14*) in which it was found that aliphatic alcohols, aldehydes, ketones, esters and amines of some aromatic compounds and heteroaromatic compounds were able to bind and this was on the whole, irreversible. In all cases, flavor levels higher than those normally applied in the flavoring of foods were used to obtain reliable data. Work was performed on humid starch where sorption was greater and inclusion compounds were formed on drying.

The presence of such inclusion compounds was established by Solms (*1*) who looked at an alternative way to measure binding by following the diffusion of the volatiles out of the starch system into which they had been incorporated. In these experiments, starch sols were prepared with starches from different sources, flavor compounds were added as ligands in different concentrations and the formation of complexes was studied. Starch has been shown to form complexes with a wide assortment of molecules representing different functional groups, molecular sizes, and polar and nonpolar molecules. In these complexes, the flavor compounds have been shown to lie in the center of the amylose helix, forming true molecular inclusion complexes. Entrapped flavor molecules have a very high stability. The practical significance of this can be seen in the loss of flavor during the aging of bread, which has been explained as an inclusion process taking place with the starch fraction, due to the retention of volatiles within these regions (*15*).

Most studies on the interaction of starch with other molecules have used gelatinized starch, as it was thought that the ungelatinized granule was relatively inert. However, it has been shown (*16*) that the granules can be penetrated by water, methanol, ethanol, 1-propanol and 1-butanol.

Inverse gas chromatography (*17*), in which the sample is used as the stationary phase in gas chromatography (GC), has been established as a means of measuring interactions between polymers and volatile solutes. Maier (*10*) used similar methodology to investigate the binding of volatile flavor compounds to foods and food components which served as the stationary phase in the chromatographic column. A disadvantage with this technique is controlling the humidity in the system, which will change due to the drying effect of the carrier gas.

In all the above experiments, the term binding has been used in its broadest sense to include adsorption, absorption, desorption, as well as chemical and physical binding. Binding as described in the present experiments could include surface phenomena, diffusion in the actual matrix or indeed diffusion through a porous medium. An equilibrium headspace system has been developed to try and address some of these problems. The method measures the overall loss of volatiles from the headspace in a closed system which includes contributions from all these phenomena. Native starch was used with a moisture content of 10 to 12%. This was present predominantly in the crystalline form interspersed with some amorphous regions.

Experimental

Methods. The wheat starch (particle size 15-40 microns), was supplied by ABR Foods Ltd. Flavor compounds 2,3-butanedione (diacetyl), benzaldehyde, ethyl acetate, hexan-1-ol, dodecane and propionic acid were obtained from Sigma Chemical Co. (Poole, UK).

Static Headspace Method. The wheat starch was placed in a 500 mL screw-top glass bottle, which was allowed to equilibrate for 30 minutes in a 25 ^{o}C water bath. A portion of the aroma compound (10 mL) was placed in a 50-mL screw-top glass bottle and sealed with a rubber septum. This was then left to equilibrate in a water bath at 25 ^{o}C to give an equilibrium headspace. The experimental temperatures were maintained by a circulating water bath. A portion of headspace (10 mL) was withdrawn from the sealed volatile bottle over a period of one minute using a gas-tight syringe. This volume was then injected into the test bottle containing the starch to produce an atmosphere with a known volatile concentration.

Gas Chromatographic Analysis of Headspace. The change in the headspace concentration was measured over time by withdrawing samples of headspace (1 mL) for GC analysis at appropriate time intervals. GC analysis of the volatile was performed on a Perkin Elmer Sigma 3B GC equipped with a flame ionization detector (FID). A polar BP20 column (25 m length x 0.33 mm, 1.0 μm film thickness, SGE, Milton Keynes, UK) was used to analyze the volatiles in the headspace sample. The GC conditions were 260 °C for the detector and injector temperatures with an isothermal oven temperature depending on the volatile under analysis: 60 °C for diacetyl; 90 °C for ethyl acetate, 1-hexanol and propionic acid; 130 °C for benzaldehyde and dodecane. Helium was used as the carrier gas at a pressure of 14 psi.

Data Handling. The peaks and peak areas were recorded by a Hewlett Packard 3392A integrator. Changes in the headspace concentration in the experimental system were expressed either as a change in the GC peak area, or on a relative percentage scale where the first sample was considered as 100%. Experiments were carried out for 1 hr and each experiment was repeated three to four times with the results expressed as arithmetic means ± standard deviation. Variation was expressed as percentage coefficient of variation (SD x 100 ÷ mean). Control experiments were performed in the absence of starch to determine the change in headspace concentration caused by the sampling regime, and these values were subtracted from the starch sample results to obtain a true value for volatile binding.

Semi-Dynamic Headspace Method. The experimental design was the same as that for the static method, except that a dumbbell-shaped PTFE magnetic stirrer (45 mm length x 9 mm) was placed at the bottom of the test bottle containing the starch. The system (solid and air) could therefore be stirred vigorously.

Results and Discussion

Static Headspace Method. The volatiles used in the experiment were chosen to represent each major functional group found in food aromas — a ketone, aldehyde, ester, alkane, alcohol and fatty acid. The binding of volatiles to starch was investigated under static conditions where it was assumed that the volatiles in the headspace were in equilibrium with the starch. Blank runs (with no starch) were also carried out and data corrected for the consumption of headspace volatiles by the sampling system. Table I shows the change in diacetyl binding in this system over a period of 1 hr when the experiment was run four times. The variability of the system is unacceptably high, with percentage coefficient of variation values (% CV) around 20%.

One of the causes of variation in this experiment was the different initial amounts of volatile introduced into the bottles (compare the different peak areas at 1 min for the four experiments). To determine the contribution of this to the overall variation, the data were expressed on a relative scale with the initial volatile amounts set to 100%. Figure 1 shows the graph of these data, but it is clear that there is still variation in the control and in the sample headspace concentrations

Table I. Starch Binding Kinetics for Diacetyl as Measured by Static Headspace GC Method

Time *(min)*	GC Peak Area (FID response) Expt 1	Expt 2	Expt 3	Expt 4	Mean	Standard Deviation	% CV
1	77361	68934	49400	60222	63979	11976	18.7
10	76515	64655	40279	50904	58088	15826	27.2
20	62566	53549	40129	39249	48873	11231	23.0
30	59393	48250	32947	35139	43932	12324	28.1
40	49790	45175	33563	33270	40449	8337	20.6
50	49005	39900	30477	40560	39985	7574	18.9
60	49562	35584	26900	37947	37520	9379	25.0

(mean % CV was 20%). The GC variability was tested and found to be low, so the explanation seemed to be that the air phase in the bottles was not thoroughly mixed and contained regions with different concentrations. However, in the samples containing 6 g and 12 g starch, binding was taking place. The rate at which this occurred was similar for the 6-g and 12-g sample. Binding was most rapid over the first twenty minutes. The effect of increasing the mass of starch does not seem to affect the initial slope of the curves significantly but, since there is variation in the control (0 g) bottle, it is difficult to interpret these results further.

Binding of Different Volatiles to Starch at 25 °C. Before attempting to improve the system, further experiments were carried out to gain some idea of the binding of other volatiles to different weights of native starch (6 and 12 g). Figure 2 shows the binding of the volatiles to 6 g starch at 25 °C, and Figure 3 shows the binding to 12 g at the same temperature. A similar pattern of binding is observed for each volatile, with 1-hexanol showing a faster rate of binding than ethyl acetate. The results for propionic acid are not represented here, as complete binding occurred in the first five minutes, making it difficult to collect sufficient data to represent this graphically. However, it is interesting to note the rapid uptake of aroma volatiles (most are bound in 10-20 min) and the extent of binding of these compounds (60 to 80% of the volatile is bound) to native starch. Maier (*14*) also showed alcohols, aldehydes and ketones were strongly bound to carbohydrate matrices.

Semi-Dynamic Headspace Method. Improvement of the original, static system was investigated by stirring the air and solid phases vigorously using an adapted magnetic stirrer. It was envisaged that such a modification might produce a homogenous air phase and thus reduce error. Results from the static and semi-dynamic systems are shown in Figures 1 and 4. The control sample (no starch present) gave more consistent results in the semi-dynamic system compared to the static system, which is probably due to the better mixing of the air and solid phases,

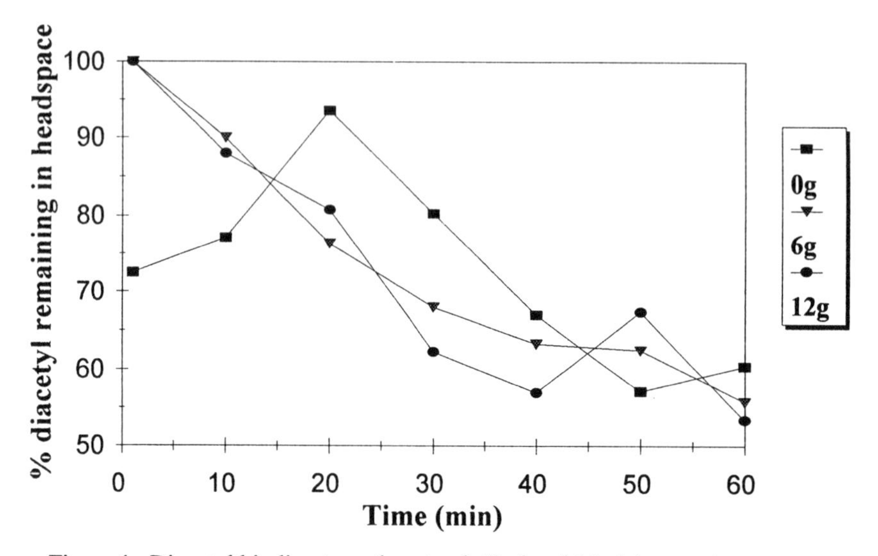

Figure 1. Diacetyl binding to native starch (0, 6 and 12 g) in a static system at 25 °C.

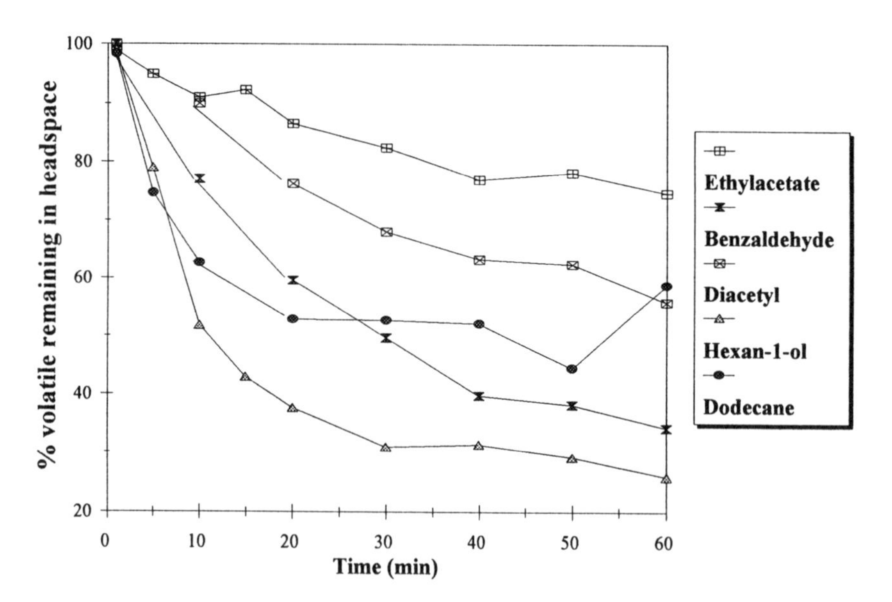

Figure 2. Binding of some aroma volatiles to native starch (6 g) at 25 °C.

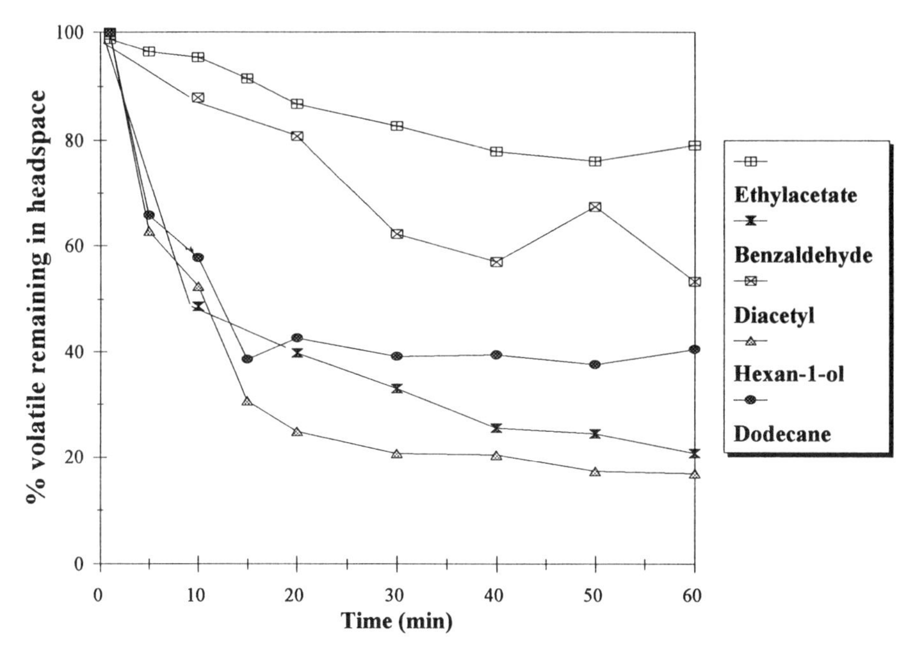

Figure 3. Binding of some aroma volatiles to native starch (12 g) at 25 °C.

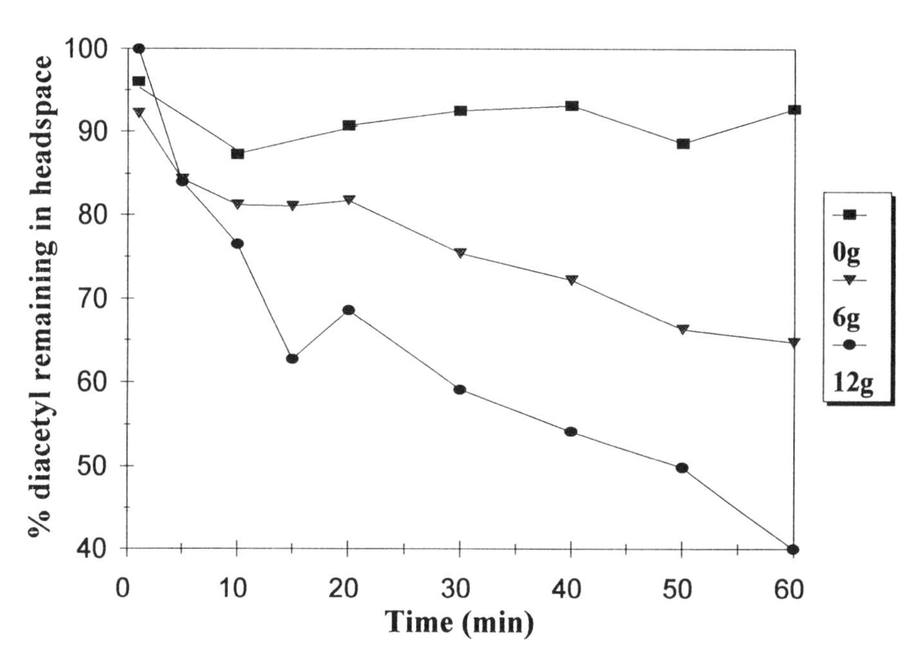

Figure 4. Diacetyl binding to native starch (0, 6 and 12 g) in the semi-dynamic system at 25 °C.

thus giving more representative and consistent headspace samples. Despite this, the two systems exhibit a similar rate of binding. On comparing the data for diacetyl binding in a static and semi-dynamic system, the effect of mixing the phases can be observed. In Table I, the % CV for the static system was between 15 and 30%, which is within the reported variation for headspace analyses, but is likely to make interpretation of binding data difficult. However, it is interesting to note that with the semi-dynamic system (Table II), the mean % CV was less than 10%, indicating that mixing improves interaction between the solid and the air resulting in less variation within the experiment.

Table II. Starch Binding Kinetics for Diacetyl as Measured by Semi-Dynamic Headspace GC Method

Time	GC Peak Area (FID response)				Mean	Standard	% CV
(min)	Expt 1	Expt 2	Expt 3	Expt 4		Deviation	
1	70938	59692	69319	61598	65387	5570	8.5
10	56638	52222	60121	61426	57601	4117	7.1
20	55259	50714	64172	53584	55932	5805	10.4
30	52213	43847	52550	48500	49277	4059	8.2
40	46935	45389	51950	44194	47117	3411	7.2
50	42567	40940	48143	41411	43265	3323	7.7
60	40507	36952	49114	42530	42276	5109	12.1

Conclusion

These preliminary studies demonstrate the feasibility of measuring the binding of a range of aroma volatiles to native starch powders. The GC analysis of the headspace gives reproducible results for the initial slopes of binding, but stirring within the sample bottle reduces the random error in measurement of binding of volatiles to starch powders. Since this chapter was written, a semi-automated method has been developed with variations below 5%, and this has been used to measure the effects of surface adsorption, partition, diffusion and physical properties on volatile binding. The availability of consistent data from binding experiments should also facilitate attempts to model some of the data to study the relative importance of processes such as diffusion in the overall binding process.

Acknowledgments

Miranda Hau wishes to thank the ACTIF2 Consortium for a studentship. This work was carried out as part of a UK Government LINK project supported by the Ministry of Agriculture, Fisheries and Food and a consortium of fourteen industrial companies.

Literature Cited

1. Solms, J.; Osman-Ismail, F.; Beyeler, M. *Can. Inst. Food Sci. Technol.* **1973,** *6,* A10-A16.
2. Franzen, K. L.; Kinsella, J. E. *J. Agric. Food Chem.* **1974,** *22,* 675-678.
3. Gremli, H. A. *J. Am. Oil Chem. Soc.* **1974,** *51,* 95A-97A.
4. Slade, L., Levine, H. In *The Glassy States in Food*; Blanshard, J. M. V. B.; Lillford, P. J., Eds.; Nottingham University Press: Nottingham, England, **1993,** pp 35-101.
5. Jin, X.; Ellis, T. S.; Karasz, F. E. *J. Polym. Sci. Polym. Phys.* **1984,** *22,* 1701-1717.
6. Kalichevsky M. T.; Jaroszkiewicz, E. M.; Ablett, S.; Blanshard, J. M. V. B. *Carboh. Polymers,* **1992,** *18,* 77-88.
7. Kalichevsky, M. T.; Jaroszkiewicz, E. M.; Blanshard, J. M. V. B. *Polymer*, **1993,** *34,* 346-358.
8. Maier, H. G. *Z. Lebensm- Unters. u. Forsch,* **1972,** *149,* 65-69.
9. Maier, H. G. *Lebensm- Wiss u. Technol,* **1972,** *5,* 1-6.
10. Maier, H. G. *J. Chromatog.* **1969,** *45,* 57-62.
11. Maier, H. G. *Deutsch. Lebensm. Rundschau,* **1974,** *10,* 349-351.
12. Maier, H. G. *Z. Lebensm. Unters. Forsch,* **1970,** *144,* 1-4.
13. Maier, H. G. *Z. Lebensm- Unters. Forsch.*, **1973,** *151,* 384-387.
14. Maier, H. G. In *Proc. Int. Symp. Aroma Research*; Maarse, H; Groenen, P. J., Eds.; Wageningen, Netherlands, **1975**, 143-157.
15. Schoch T. J. *Bakers Digest,* **1965,** *39,* 48-50.
16. Bemiller, J. N.; Pratt, G. W. *Cereal Chem.* **1981,** *58,* 517-520.
17. Gilbert, S. G. *Adv. in Chromatogr.*, **1984,** *23,* 199-228.

Chapter 11

Interactions Between Pectins and Flavor Compounds in Strawberry Jam

Elisabeth Guichard

Laboratoire de Recherches sur les Arômes, Institut National de la Recherche Agronomique, 17 rue Sully, 21034 Dijon Cedex, France

Gelling agents are added to commercial products to achieve desired firmness or consistency. These agents should not interfere with the aroma, flavor or taste of the product to which they are added. Among them, pectic substances find many applications, particularly in jam manufacturing. Composition of headspace, consistency, taste and flavor characteristics were determined in jam made with different pectins. At the usual concentrations, high methoxylated pectin induced an undesirable modification of typical flavor and intensity of flavor and taste, whereas low methoxylated pectin induced few alterations. At fixed concentration and molecular weight, a decrease in degree of esterification produced a significant decrease in consistency and noticeable modifications of the flavor perception and headspace composition, but no taste alteration. Mechanical reduction of pectin molecular weight significantly modified only the consistency.

Texture characteristics, and particularly consistency, are important factors in the overall acceptability of jam. Pectin is a naturally occurring polysaccharide, mainly extracted from citrus peel and apple pomade. High methoxylated pectins (HMP) are used to form gels in acidic media of high sugar content (*1*), and low methoxylated pectins (LMP) are used in products of lower sugar content. The strength of gels obtained with LMP varies essentially with concentration of calcium ions in the medium but also with the molecular characteristics of the polysaccharide. At a specific degree of methylation, the physical properties of a HMP are modified by the distribution and location of the remaining free carboxylic groups (*2*). The molecular weight of pectin can also influence some gel strength characteristics. Crandall and Wicker (*2*) found that the elasticity modulus was influenced primarily by the short, rigid chains and was independent of pectin molecular weight (MW). On the contrary, they also concluded that breaking

0097–6156/96/0633–0118$15.00/0

strength was influenced primarily by the longer, more flexible chains which remained cross-linked after the shorter, rigid chains had ruptured, (which were related to MW). Panchev *et al.* (*3*) tested different pectins and found that the optimal strength of the gel corresponded to degree of esterification (DE) values between 57–58 %. However, they also tested pectins in which the MW was decreasing with degree of esterification, thus precluding a definitive conclusion about the relation between gel strength and DE.

Many studies have demonstrated that hydrocolloids not only modified viscosity, but often reduced intensities of odor, taste and flavor (*4, 5*). Some evidence indicated this masking effect varied with the type and concentration of hydrocolloid used. Most of these studies, such as the one by Marshall and Vaisey (*6*), concerned the effect of hydrocolloids on taste qualities in model solutions. Lundgren *et al.* (*7*) investigated the effect of pectin on odor, taste and flavor intensities in jams, but at concentrations 10-times higher than those used in jam manufacturing. The objective of our study was to clarify the influence of the amount of pectin added, and the DE and MW of that pectin on sensory characteristics (such as consistency of the gel, typical flavor character and intensity of flavor), and on amounts of volatile compounds in headspace.

Experimental Procedures

Pectin Preparation. One rapid-set HMP and one LMP from Mero Rousselot Satia (France) were used. Both were non-standardized citrus pectins (245° SAG for the HMP), currently recommended for standard jam manufacturing (60° Brix, 45% fruit).

Experimental Samples. Four experiments were carried out using the following design, as shown in Table I:

1. Jams with increasing amounts of HMP at 83% DE (0, 0.05, 0.1, 0.2 and 0.4 %)
2. Jams with increasing levels of LMP at 37% DE (0, 0.1, 0.2, 0.4 and 0.6 %).
3. Jams with 0.2% of pectin at varying degrees of esterification (HM pectin was de-esterified according to Guichard *et al.* (*8*)), giving three pectins with degrees of esterification of 83, 66 and 54%).
4. Jams with 0.2% of pectin with varying molecular weights (molecular weight of HM pectin was reduced to 86.000, 75.000, 59.000 and 32.000).

In each, a control sample without added pectin was included. The jam preparation has been previously described (*8*).

Chemical Analysis

Isolation of Volatiles. A headspace analysis was used in order to avoid gel disruption. Four hundred grams of jam were introduced into a 1-L flask and extracted according to the method described by Guichard and Ducruet (*9*): the vapor phase was stripped for 19 hr by a stream of 110 mL/min nitrogen, and the volatile

Table I. Characteristics of Pectins Used in Different Experiments

Experiment No.	*Type of Pectin*	*Degree of Esterification (%)*	*Molecular Weight*
1	H M	83	86.000
2	L M	37	59.000
3	H M	83	86.000
		66	86.000
		54	86.000
4	H M	83	86.000
		83	75.000
		83	59.000
		83	32.000

compounds were trapped in a liquid-liquid continuous extractor containing 250 mL of a 10% ethanolic solution, and continuously extracted with 100 mL of dichloromethane. Each analysis was performed in duplicate. For quantification of volatiles, *n*-tridecane (25 pg/g of jam) was added as an internal standard in the solvent extract.

Gas Chromatography. Gas-chromatographic analyses of the extracts were performed using a Girdel 300 gas chromatograph equipped with a chemically bonded DB-5 fused silica capillary column (30 m, 0.32 mm i.d., 1 µm, J & W Scientific Inc.). The injection temperature was 220 °C and detector, 250 °C. Extracts (1µL) were injected splitless. After injection, the oven temperature was held at 30 °C for 5 min and then programmed at 2 °C/min to 220 °C. The flow rate of the carrier gas (H_2) was 37 cm/s. For quantification, an Enica 10 integrator (Delsi France) was used. Odors of compounds eluting from the column were assessed by three judges (*10*).

Gas Chromatography–Mass Spectrometry. Compound identifications on each extract were performed using a Nermag R 10-10/C mass spectrometer coupled with the gas chromatograph described above, and equipped with a DB-5 column (60 m, 0.32 mm i.d; 1 µm, J & W Scientific Inc.). Ionization was by electronic impact at 70 eV.

Sensory Analysis

Subjects. Eighteen subjects were selected on their ability to memorize and recognize basic tastes and odors, and to rank jams with different pectin levels on oral consistency. Quantitative descriptive analysis was performed on the jams. During four sessions, descriptive terms were generated by the judges from individual evaluation of commercial strawberry jams. During a fifth session, panelists rated the intensity of each term using an unstructured, 13-cm scale. Results were then discussed in order to establish a final list of ten flavor attributes: total intensity, typical flavor, fresh strawberry, unripe strawberry, overripe strawberry, cooked strawberry, candied fruit, caramel, artificial, lemon.

Evaluation of Consistency. The most preferred level of jam consistency was estimated by each member of the panel. This was indicated as "just right" in the center, with anchors of "not hard enough" and "too hard" at the opposite ends of the scale. Evaluation of consistency was made separately by the same group of subjects. Oral consistency was rated using an unstructured, 130-mm scale, with a verbal anchor point at each end (left anchor = very soft; right anchor = very hard), as already described (*11*).

Results and Discussion

Ideal Consistency. A histogram of the ideal values (Figure 1) showed that only one subject preferred jam with a HMP concentration higher than 0.2%. The mean ideal consistency was calculated to correspond to a HMP concentration near 0.11%. This value seemed low compared to amounts currently used in jams (0.2%) but since the HMP was not standardized, the corresponding amount of standardized pectin (150° SAG, instead of 245°) should be 0.18%.

Evaluation of Oral Consistency with Pectin Concentration. At the same concentration, HM pectin gives a harder gel than the LM pectin (Figure 2) and the oral consistency of jam made with 0.6% of LM pectin is equivalent to that made with 0.25% of HM pectin (60° Brix). At a same pectin level, the oral consistency of jam increases with the Brix level.

For the same level of pectin (0.2%), the oral consistency increases proportionally with degree of esterification and molecular weight (Figure 3).

Volatile Flavor Compound Identifications. Fifteen key volatile compounds of flavor significance were identified in strawberry jam by gas chromatography–mass spectrometry, and their corresponding aroma descriptors are listed in Table II. Changes in levels of these compounds were compared against sensory differences among the jam samples for the various types of pectins.

Influence of Amount of HM Pectin on Jam. Figure 4 shows that the overall intensity and typical flavor notes of the jam decreased with higher amounts of HM pectin. This could be explained by a decrease of strawberry and caramel notes and an increase of the candied note. Headspace analysis showed that only six of the compounds analyzed were significantly affected by an increase in pectin. Figure 5 shows that adding only 0.05% of pectin drastically decreased the amount of ethyl hexanoate, and to a lesser extent the other compounds. A decrease in the headspace

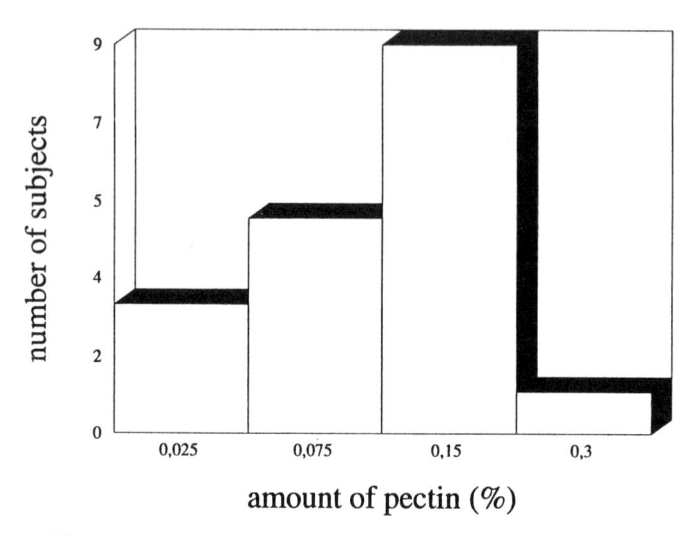

Figure 1. Histogram of the ideal consistency values of jam.

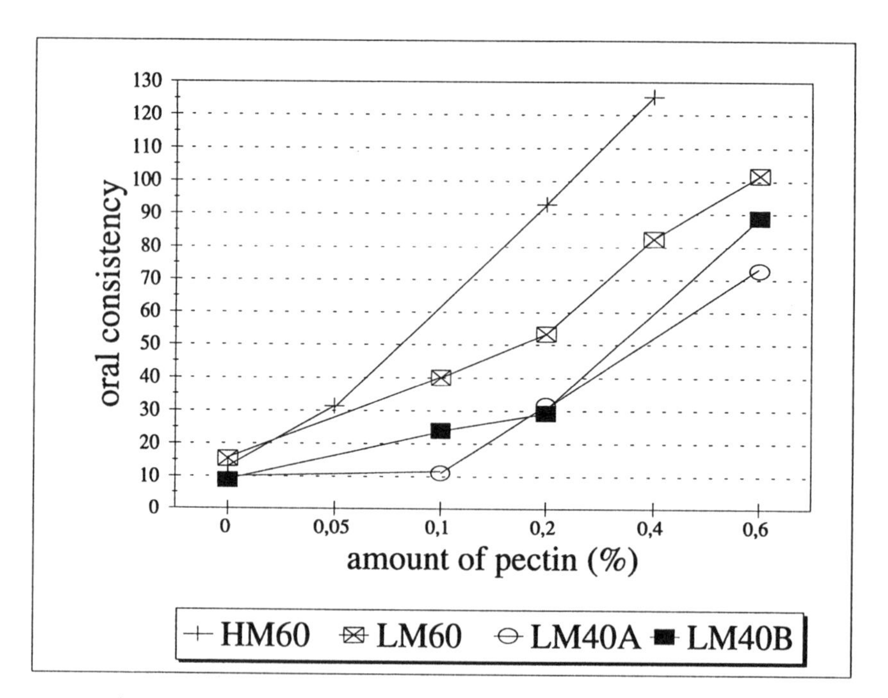

Figure 2. Change in oral consistency with pectin concentration.

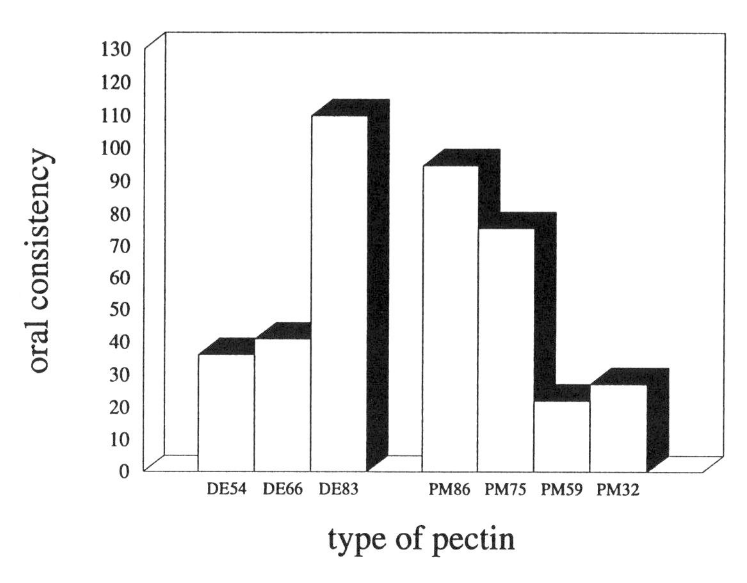

Figure 3. Oral consistency as a function of the degree of esterification and molecular weight of pectin, for the same amount of pectin (0.2%).

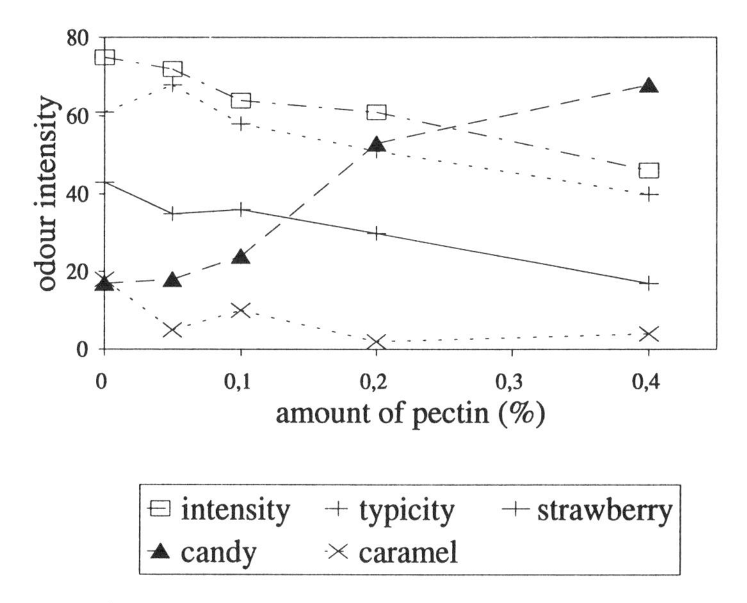

Figure 4. Effect of the amount of high methoxylated pectin on the sensory characteristics of the jam.

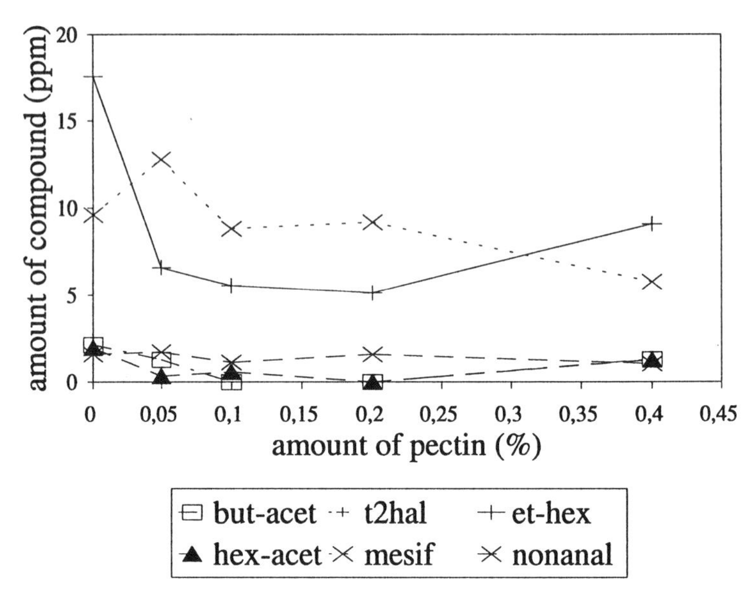

Figure 5. Effect of the amount of high methoxylated pectin on the amount of volatile compounds in the jam.

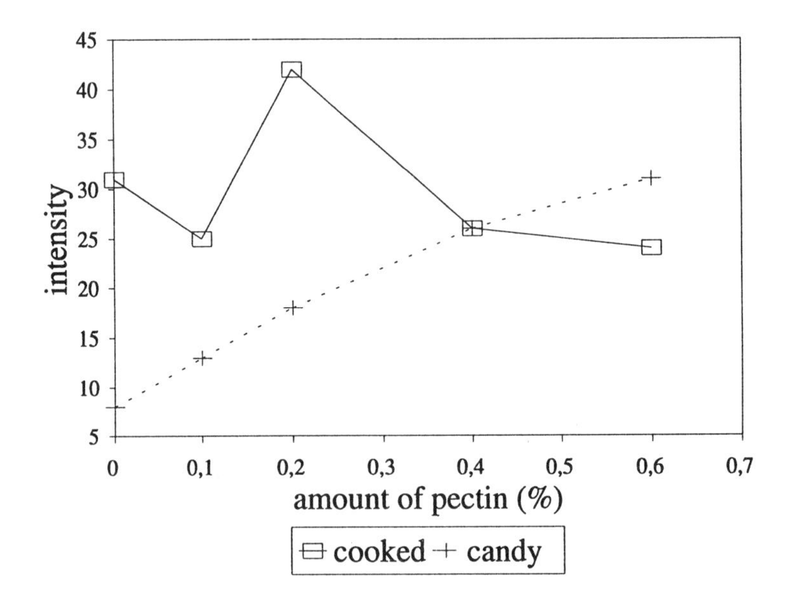

Figure 6. Effect of the amount of low methoxylated pectin on the sensory characteristics of the jam.

of mesifurane (2,5-dimethyl-4-methoxy-2,3-dihydrofuran-3-one), a compound described as caramel-like, could partly explain the decrease of the caramel note in the jam, and also the typical jam flavor. Moreover, the amounts of some esters such as butyl acetate, ethyl hexanoate and hexyl acetate, (compounds exhibiting flowery or fruity notes), and aldehydes such as nonanal and *trans*-2-hexenal, (green odors),

Table II. Volatile Compounds Showing Significant Differences Among Jam Samples

Code	*Flavor Compound*	*Odor*
met-but	Methyl butanoate	Fruity
pent3on2	3-Penten-2-one	Herbaceous
et-but	Ethyl butanoate	Fruity
but-acet	Butyl acetate	Flowery
fur	Furfural	Pungent,caramel
t2hal	*trans*-2-Hexenal	Fresh green
met-hex	Methyl hexanoate	Fruity
bzal	Benzaldehyde	Almond
hex-acid	Hexanoic acid	Cheese
et-hex	Ethyl hexanoate	Fruity
octal	Octanal	Potato
hex-acet	Hexyl acetate	Fruity
mesif	Mesifurane [a]	Caramel
nonanal	Nonanal	Flowery
oct-acid	Octanoic acid	Cheese

[a] 2,3-Dihydro-2,5-dimethyl-4-methoxy-3(H)-furan-3-one

decreased with the typical aroma of the jam, from 0 to 0.2% of pectin. A small increase in amounts of these compounds in the jam prepared with 0.4% of pectin could be explained by the greater exchange area in the flask during headspace analysis. When the jam was poured into the flask, it was too thick to spread out. Since Douillard and Guichard (*12, 13*) demonstrated the contribution of these compounds to strawberry aroma, the variations in concentration could be directly responsible for the modifications of aroma.

Influence of the Amount of LM Pectin on Jam. The sensory assessment of the jam made with different amounts of LMP did not show significant differences for typical flavor and intensity of the aroma. As shown in Figure 6, the cooked fruit note varied independently of the amount of pectin and the candied fruit note was the only flavor characteristic which increased with it, although to a lesser extent than

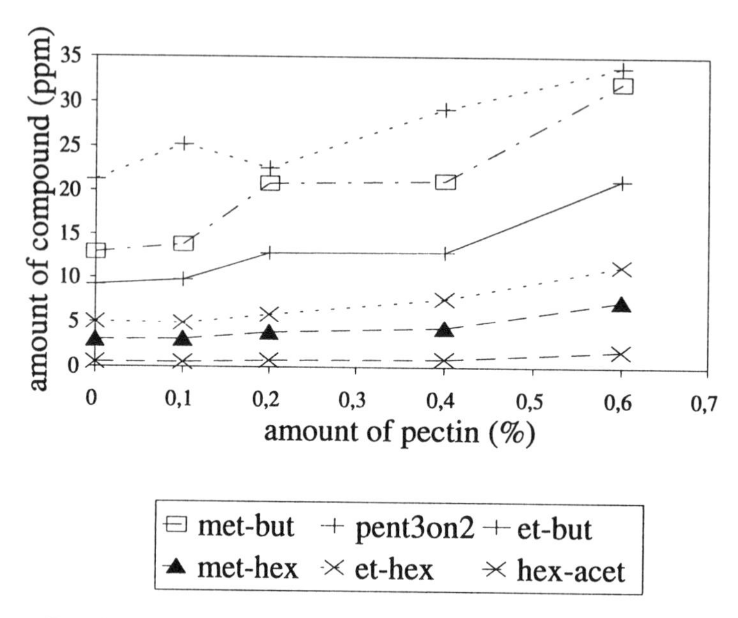

Figure 7. Effect of the amount of low methoxylated pectin on the amount of volatile compounds in the jam.

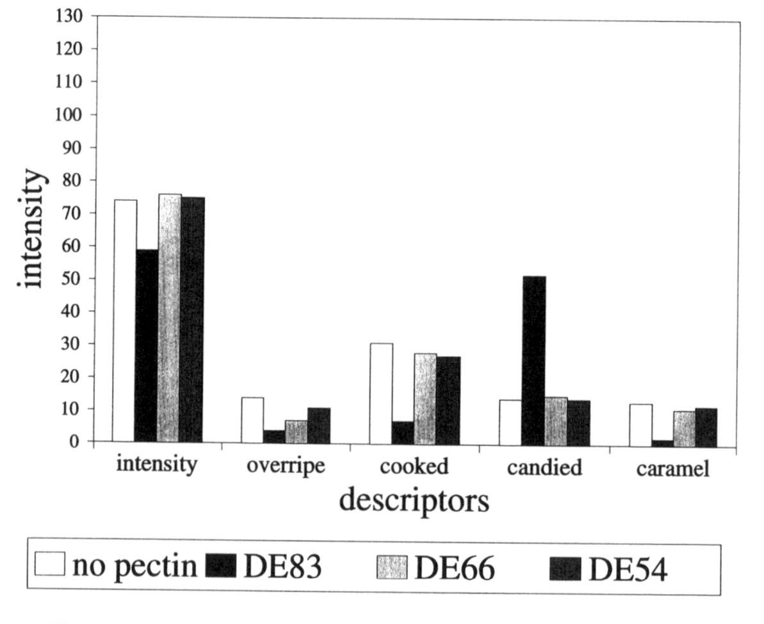

Figure 8. Effect of pectin degree of esterification on the sensory characteristics of the jam.

when using HMP. In the headspace, a progressive increase of concentration of such esters as methyl- and ethyl butanoates, methyl- and ethyl hexanoates and hexyl acetate, which possess fruity notes, and of 3-penten-2-one, which is described as herbaceous, could be observed (Figure 7). These changes in amounts did not seem to affect the overall perception of aroma. Nevertheless, it was notable that volatiles increased when LMP was added instead of the decrease observed when HMP was added. The enhancement of the candied note could not be explained by the increase of these substances. It was more probably due to the influence of some other volatile compounds which we could not detect instrumentally.

Influence of Degree of Esterification. Aroma characteristics differed between the control and the jams in which the original pectin were added (Figure 8). As before, the overall intensity of aroma, the cooked fruit and the caramel notes were found to decrease while the candied fruit note increased. However, decreasing the DE of the pectin seemed to restore the original intensities of overall aroma and its characteristics.

The amounts of nonanal, methyl- and ethyl butanoates and hexanoic acid were lower in the headspace of the jam made with original pectin (Figure 9). This could explain the lower aroma intensity perceived in that sample, compared to the jam made without pectin. Concentrations of many volatile compounds in the jam were affected by modification of DE. The higher aroma intensity reported when DE was decreased from 83 to 54 could also be related to the increase in amounts of these compounds, to a level near that found in the jam without pectin. On the contrary, the higher caramel and cooked notes found when using low DE pectins could be explained by higher furfural amounts (Table II).

Influence of Molecular Weight. As in the other experiments, addition of the original pectin caused an increase in the candied fruit note and a decrease in the caramel note. Moreover, the fresh strawberry note was lower in the reference as opposed to the overripe strawberry note (Figure 10). Volatile aromatic compounds, particularly esters, ketones and some aldehydes responsible for fruity and fresh notes, were found at lower amounts in these jams. However, the overripe strawberry note found by the panelists could not be related to a significantly higher amount of any detected compounds. Moreover, the addition of pectin, whatever the molecular weight, seemed to reduce the perception of the overripe note and to increase the perception of the strawberry note. However, no significant difference was observed among the jams in the headspace analysis. It was thus impossible, in this last experiment, to explain the sensory assessment of the jam based on the headspace volatile results.

Conclusions

Our results showed that addition of pectin not only modified the oral consistency of jam, but also caused a decrease in both taste and flavor intensities, thus confirming previous studies. For a specified pectin, this masking effect increased with the level of pectin and, at a specified level, depended on the type of pectin. Since the gelling

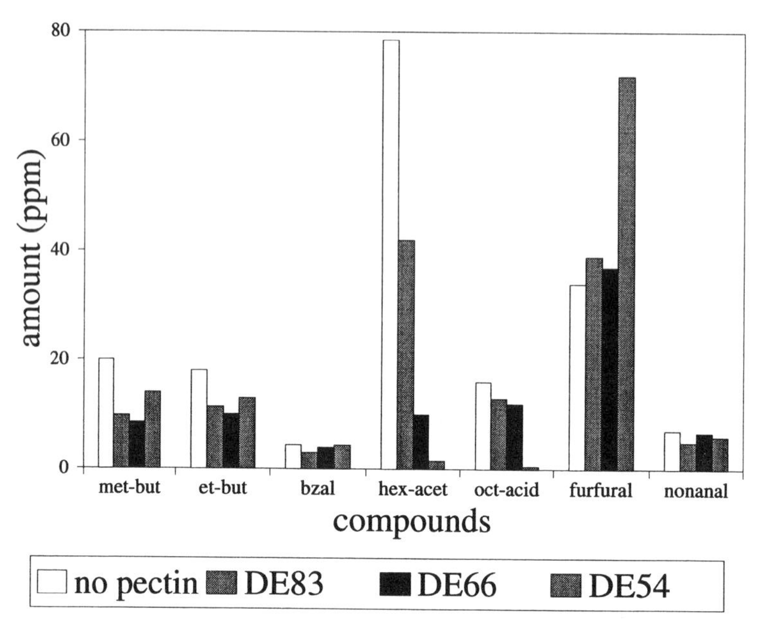

Figure 9. Effect of pectin degree of esterification on the amount of volatile compounds in the jam.

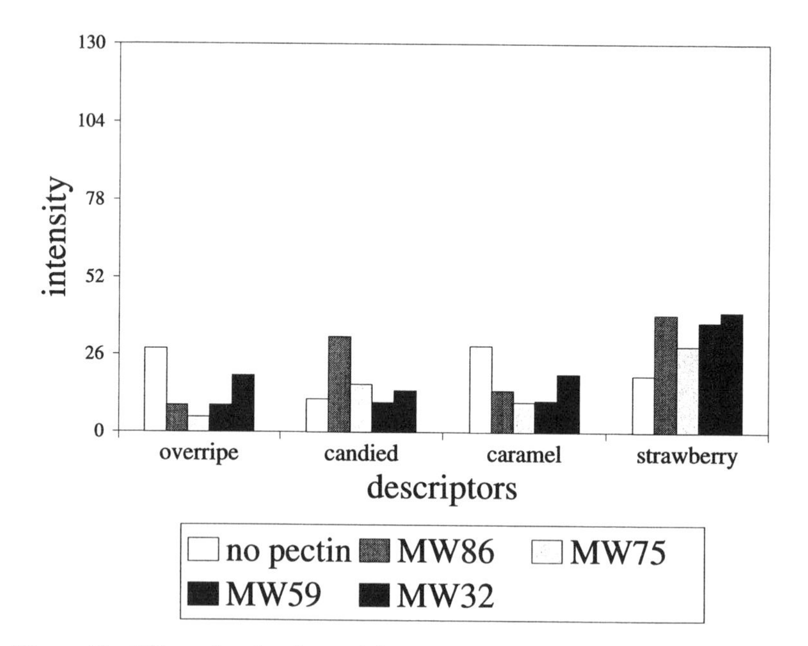

Figure 10. Effect of molecular weight of pectin on the sensory characteristics of the jam.

capacity changed with the type of pectin, this masking effect could be due only to an increase in consistency. However, at a similar level of consistency, taste intensity was reduced less with a LMP than with a HMP, demonstrating that consistency was not the only factor responsible for variations in taste intensity.

The same pattern was observed for flavor intensity. Increases in the candied fruit note with increasing consistency could be due to the observed decrease of the amount of some volatiles responsible for the fresh strawberry note in the headspace. Subjects could have used this term to describe not only a specific flavor characteristic but a joint effect of a harder consistency and a lower fresh fruit intensity.

The amounts of many volatile compounds isolated from the different jams were not significantly different. However, due to the isolation procedure used, several important key components of jam aroma were not determined. This method did not allow, for example, the detection of polar compounds such as furaneol and lactones.

Demethylation of the pectin reduced the interactions between pectin and the volatile compounds, thus allowing a better perception of the aroma. However, demethylated pectins gave a less gellified jam, the consistency of which allowed more exchanges between the aqueous phase and the vapor phase. It is thus difficult, from this experiment, to dissociate the two effects and to conclude there was a direct effect from the degree of methylation.

Literature Cited

1. Leroux, H.; Schubert, E. *Ind. Aliment. Agric.* **1983**, 615-618.
2. Crandall, P. G.; Wicker, L. In "*Chemistry and Function of Pectins*," Fishman, Jen, J. J., Eds.; ACS Symposium Series: Washington D.C., 1986, 88-102.
3. Panchev, I. N.; Kirtchev, N. A.; Kratchanov, C. G.; Proichev, T. *Carbohydrate Polymers* **1988**, *8*, 257-269.
4. Moskowitz, H. R.; Arabie, P. *J. Texture Studies* **1970**, *1*, 502-510
5. Pangborn, R. M.; Trabue, I. M.; Szczesniak, A. S. *J. Texture Studies* **1973**, *4*, 224-232.
6. Marshall, S.; Vaisey, M. *J. Texture Studies* **1972**, *3*, 173-180.
7. Lundgren, B.; Pangborn, R. M.; Daget, N.; Yoshida, M.; Laing, D. G.; McBride, R. L.; Griffiths, N.; Hyvönen, L.; Sauvageot, F.; Paulus, K.; Barylko-Pikielna, N. *Lebens. -Wiss. u.-Technol.* **1986**, *19*, 66-76.
8. Guichard, E.; Ducruet, V. *J. Agric. Food Chem.* **1984**, *32*, 838-840.
9. Guichard, E.; Issanchou, S.; Descourvières, A.; Etiévant P. *J. Food Sci.*, **1991**, *56* (6), 1621-1627.
10. Etievant, P. X.; Issanchou, S. N.; Bayonove, C. L. *J. Sci. Food Agric.* **1983**, *34*, 497-504.
11. Issanchou, S.; Maingonnat, J. F.; Guichard, E.; Etievant, P. X. *Sci. Aliments* **1991**, *11*, 85-98.
12. Douillard, C.; Guichard, E. *Sci. Aliments* **1989**, *9*, 61-83.
13. Douillard, C.; Guichard, E. *J. Sci. Food Agric.* **1990**, *50*, 517-531.

Chapter 12

Taste Interactions of Sweet and Bitter Compounds

D. Eric Walters[1] and Glenn Roy[2]

[1]Department of Biological Chemistry, Finch University of Health Sciences, The Chicago Medical School, 3333 Green Bay Road, North Chicago, IL 60064
[2]Pepsi-Cola Company, 100 Stevens Avenue, Valhalla, NY 10595

Studies of sweet and bitter taste mechanisms and sweet and bitter compounds indicate that there may exist some relationship between sweet taste receptors and bitter taste receptors. These studies include the discovery of sweet taste inhibitors which also block some bitter tastes. We describe the interactions between sweet and bitter tastes, as well as interactions of sweet and bitter compounds with food components and with food additives intended to inhibit sweet or bitter taste.

Sweet taste and bitter tastes are at the same time very different and very closely related. Sweet taste is usually considered a desirable quality, while bitter taste is usually (but not always) considered undesirable. Sweet and bitter tastes apparently have great similarity in terms of the cellular mechanisms by which they are detected. Interaction between sweet and bitter tastes is common in food systems. In food and pharmaceutical formulations, sweetness is often used to mask bitter taste. Conversely, bitterness is sometimes used to balance high levels of sweetness. If we can understand the mechanisms by which sweet and bitter are perceived, and the ways in which these mechanisms are related, we can understand and utilize relationships between sweet and bitter tastes in formulating food and pharmaceutical systems.

In this chapter, we first examine experimental results which indicate a relationship between sweet and bitter taste receptor systems. Second, we review the current extent of our knowledge about mechanisms for sweet and bitter taste perception. Next, we propose a molecular basis by which this knowledge and evidence can be rationalized. Finally, we explore the implications of the proposed rationale for controlling perception of sweet and bitter tastes in food and pharmaceutical formulations.

0097–6156/96/0633–0130$15.00/0

Experimental Results

Are There Receptors for Sweet and Bitter Taste? Much of the evidence connecting sweet and bitter tastes is empirical in nature. In spite of extensive efforts over a period of many years, no taste receptor protein has yet been isolated. One might even question whether there is such a thing as a sweet receptor since the natural ligands for such a receptor (sugars) produce taste only at concentrations greater than about 0.1 molar. In contrast, most biological receptors show affinity for their ligands at 10^{-6} to 10^{-12} molar concentrations. Perhaps sweet taste is not receptor mediated!

On the other hand, there are now many high-potency sweeteners known, including compounds which are agonists (able to induce a response) at concentrations as low as 10^{-7} molar (*1*). In addition, some sweeteners are found to act as "partial agonists" (Figure 1). That is, they exhibit activity at low concentrations, but as the concentration is increased, they are not ever able to elicit a full response. For instance, DuBois et al. showed that a number of sweeteners are unable at any concentration to match the sweetness of 10% sucrose (*2*), even though they have high potency relative to sucrose at threshold levels. This partial agonist effect is seen in many receptors of pharmacological interest, indicating that sweetness may be receptor-mediated at least for high-potency sweeteners. [Dr. Terry Acree of Cornell University makes the useful distinction between the term "high-potency sweetener" (active at low concentration) and the term "high-intensity sweetener" (able to induce a high level of sweetness).]

There are bitter compounds which are taste-active at very low concentrations. Denatonium benzoate can be detected at a concentration of 2 x 10^{-9} molar (*3*), indicating a likelihood that there are specific receptors involved. In addition, Whitney and coworkers have found a gene which controls sensitivity to the bitter taste of sucrose octaacetate (*4*).

Are There Multiple Receptor Types? The chemist is immediately struck by the broad diversity of compounds which taste sweet. The following list is representative, not exhaustive:

- sugars: sucrose, fructose, glucose; trichlorogalactosucrose (*5*)
- peptides: aspartame (*6*), alitame (*7*)
- sulfamates: cyclamate (*8*)
- heterocycles: saccharin (*9*), acesulfame (*10*)
- ureas: dulcin (*11*), suosan (*12*), superaspartame (*1*), β-aminoacid-ureas (*13*)
- arylguanidines: sucrononic acid (*1*)
- proteins: thaumatin (*14*), monellin (*15*)
- oximes: perillartine (*16*)
- terpenes: hernandulcin (*17*)
- dihydrochalcones: neohesperidin DHC (*18*)

Besides the diversity of structures, there is diversity in the quality of sweet taste associated with different compounds. Saccharin and acesulfame have a bitter-metallic aftertaste for many people, and cyclamate has a salty taste at higher

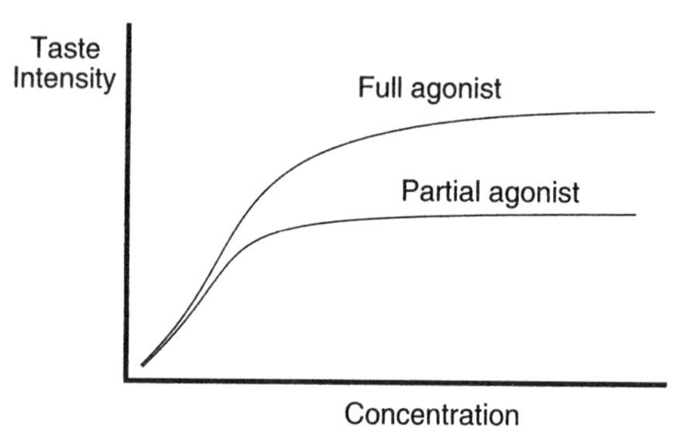

Figure 1. Illustration of the "partial agonist" phenomenon.

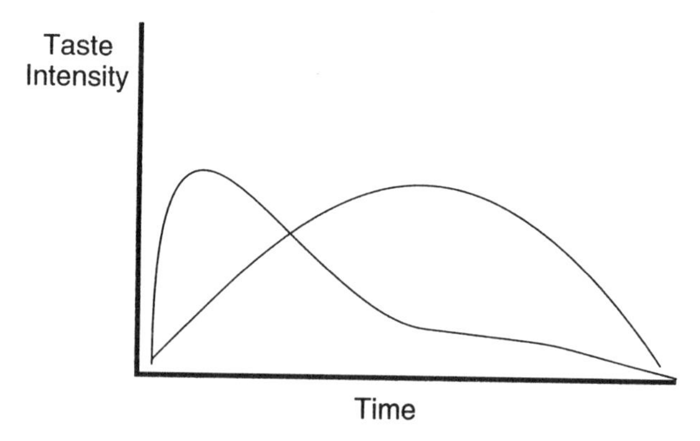

Figure 2. Schematic representation of rapid and slow temporal profiles.

concentrations. If receptors are involved in sweet taste transduction, we must seriously consider the possibility of *multiple receptors*.

Similarly, there is a tremendous diversity of compounds which taste bitter: alkaloids (quinine, strychnine), acyl sugars (sucrose octaacetate), peptides, heterocycles (caffeine), ureas, arylguanidines, oximes (*19*), terpenes, dihydrochalcones (*18*), and countless others are known. Bitter taste is not as thoroughly studied as sweet taste, but it seems likely that there are different taste qualities which we lump together under the heading of bitter. We simply do not have an extensive vocabulary for describing bitter tastes. Since the gene controlling response to sucrose octaacetate does not control response to all bitter tastes (*4*), we must again expect that there are multiple receptor types involved.

There is yet another indication of the likely existence of multiple receptor types. Blends of sweeteners may produce *synergy*, a higher level of sweetness than would be predicted if the sweetness of the components were simply additive (*20*). A single receptor type would not be likely to produce such an effect. In fact, blends of sweeteners which are structurally similar (e.g., saccharin + acesulfame) do *not* produce synergy, probably because they are competing for a single receptor type. There appear to be no studies indicating whether synergy occurs with bitter-tasting substances.

Are There Multiple Mechanisms for Sweet and Bitter Taste Transduction? Variations in temporal profile (*20*) may be another clue that there is more than a single way in which sweetness is perceived. Some sweeteners have a rapid onset and their sweet taste clears quickly, while others have slow onset and lingering sweet taste (Figure 2). This may be an indication that there are not just multiple receptors, but multiple mechanisms of transduction, some which respond rapidly and others which give a prolonged response.

Are Sweet and Bitter Receptors Related? Most intriguing is the frequent observation that small structural changes can convert potently sweet compounds into potently bitter compounds (selected examples are shown in Table I). Stereoisomers of aspartame are bitter, as are simple analogs of dulcin, acesulfame, and sucralose. Glycine and many D-amino acids are sweet, while many hydrophobic L-amino acids are bitter. Sugar esters such as sucrose octaacetate are bitter as well. Conversely, when a small structural change was made in bitter components of citrus fruits, the resulting dihydrochalcones were found to be quite sweet.

The existence of many structures which are both sweet and bitter indicates that some sweet receptors and some bitter receptors may have similar binding sites. In fact, we frequently encounter homologous series which contain sweet, bitter, and sweet + bitter tasting compounds. The following example from Cohn (*21*) is typical and suggestive of the degree of similarity which may exist between sweet and bitter taste receptors:

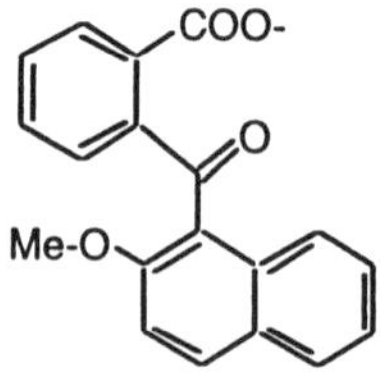

COO-
O
Me-O
sweet

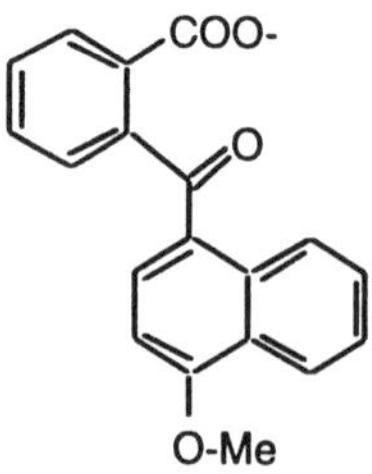

COO-
O
O-Me
bitter

COO-
O
O-Me
sweet + bitter

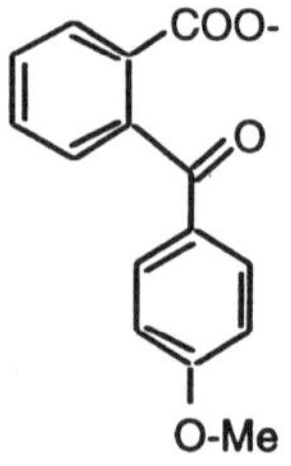

Table I. Structural Comparison of Sweet Compounds and Bitter Analogs

Sweet compound	Bitter compound	Ref.
L-Aspartyl-L-phenylalanine methyl ester (aspartame)	L-Aspartyl-D-phenylalanine methyl ester	6
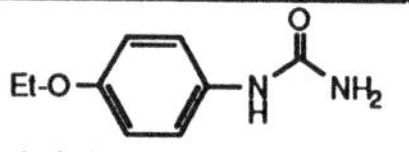 Et-O–C₆H₄–NH–C(=O)–NH₂ dulcin	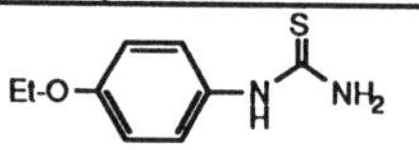 Et-O–C₆H₄–NH–C(=S)–NH₂	11, 21
sucrose	sucrose octaacetate	22
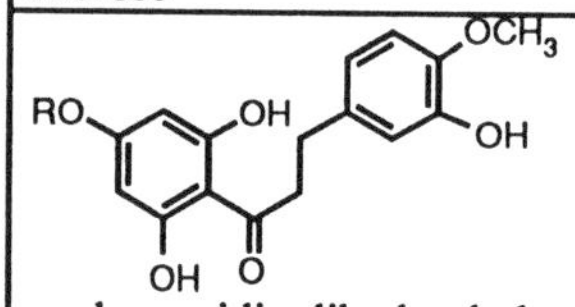 neohesperidin dihydrochalcone (R = β-neohesperidosyl)	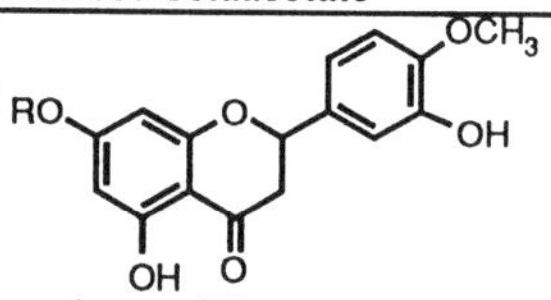 neohesperidin (R = β-neohesperidosyl)	18
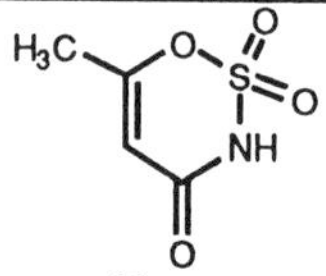 H_3C … acesulfame	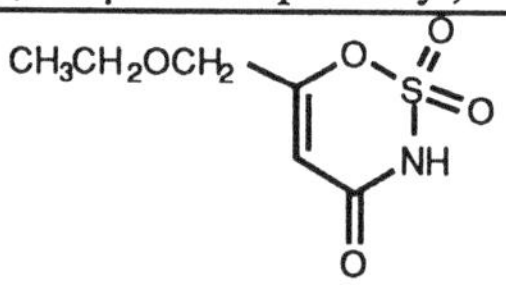 $CH_3CH_2OCH_2$ …	10, 23
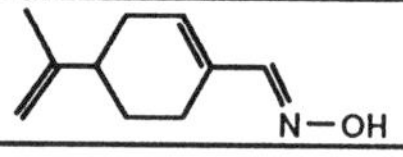 N—OH	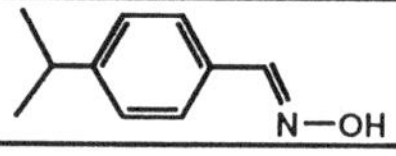 N—OH	24

The structural diversity found among sweet tasting compounds is a strong argument in favor of the existence of multiple receptor types for sweet taste--how could a single receptor respond to so many different structures? And yet all sweet taste *inhibitors* discovered to date are non-specific. The sulfonic acid-arylurea compound described by Muller et al. (*25*) blocked the sweet taste of 10 different structural types of sweeteners. In addition, it blocked some (but not all) bitter tasting compounds. This compound does not block sour or salty taste, ruling out a general mechanism such as a local anesthetic effect. These results cloud the picture regarding multiple receptor types for sweet taste, and again point to some sort of relationship between sweet and bitter taste.

McBurney and Bartoshuk have shown that, after tasting a sample of sucrose, water has a bitter taste; conversely, after tasting a sample of quinine, urea, or caffeine, water has a sweet taste (*26*). Again, we are led to suspect that there is a relationship between sweet and bitter taste mechanisms.

Current Understanding of Bitter and Sweet Taste Mechanisms

Mechanistic studies of taste transduction, often using molecular biology methods to clone and characterize important proteins, have begun to show us the details involved in sweet and bitter taste transduction (*27*). It is apparent that both tastes utilize G protein coupled receptors (GPCR), a large family of receptors which includes rhodopsin, olfactory receptors, many neurotransmitter receptors, and peptide hormone receptors. The GPCR family must be a very ancient family of proteins, since they serve a chemosensory function in some single cell organisms. For example, yeast mating factor receptors are GPCRs (*28*). The general way the GPCR system functions is as follows. The receptor protein is embedded in the cell membrane, with seven alpha-helical segments traversing the membrane. The helices pack together to form a ligand binding site. When extracellular ligand binds to the receptor, a conformational change occurs which enables the intracellular segments of the receptor to interact with a GTP-binding protein (G protein), activating the G protein. The activated G protein then activates a second messenger inside the cell; the second messenger may be adenylyl cyclase (producing cAMP, Figure 3), or a phosphodiesterase (hydrolyzing cGMP or cAMP, Figure 4) or phospholipase C (producing inositol trisphosphate and releasing intracellular Ca++ stores, Figure 5). The cAMP, cGMP, or Ca++ then may modulate the activity of an ion channel, causing the cell to depolarize or hyperpolarize.

Resting taste cells have a negative electrical potential (net negative charge inside the cell). When the potential is raised to a threshold level, ion channels in the cell membrane open and the cell briefly depolarizes (goes toward neutral). This triggers the electrical signal in the attached nerve cells.

There are numerous implementations of the G protein coupled receptor-second messenger-ion channel theme (*29*). A single receptor may activate one or more different G proteins. Multiple receptor types may activate a given G protein. One or more second messengers may be involved, and one or more types of ion channel may be affected.

Components of G protein coupled receptor systems have been found in taste cells. McLaughlin et al. have identified transducins (G proteins which had previously been found only in retinal cells) and a related G protein which they named gustducin (*30*). In the retina, transducins couple the receptor rhodopsin to a phosphodiesterase which hydrolyzes cyclic GMP. Transducins and gustducin have sequence similarity both in the receptor binding domain and in the phosphodiesterase activation site. The bitter compound denatonium was found to

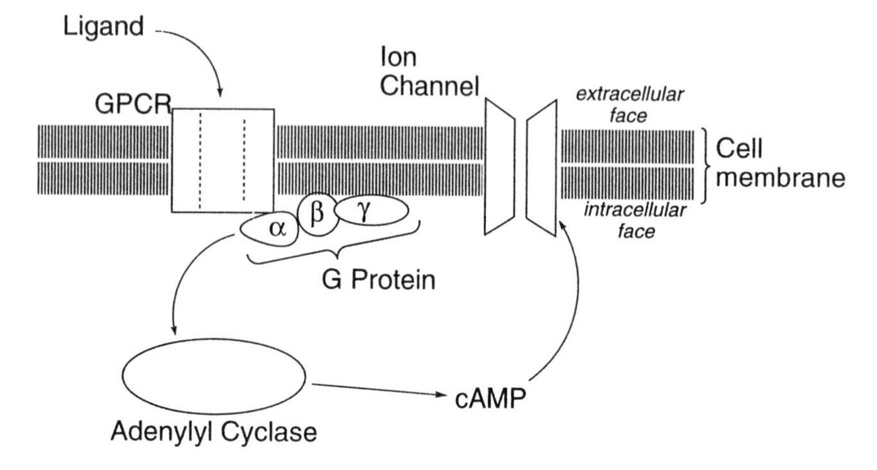

Figure 3. G Protein coupled receptor system in which the second messenger is adenylyl cyclase/cyclic AMP.

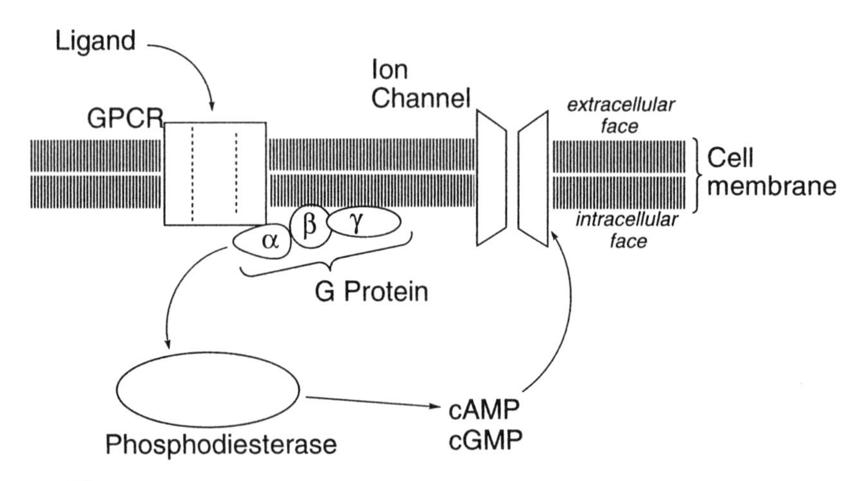

Figure 4. G Protein coupled receptor system in which the second messenger is a phosphodiesterase which degrades cyclic AMP or cyclic GMP.

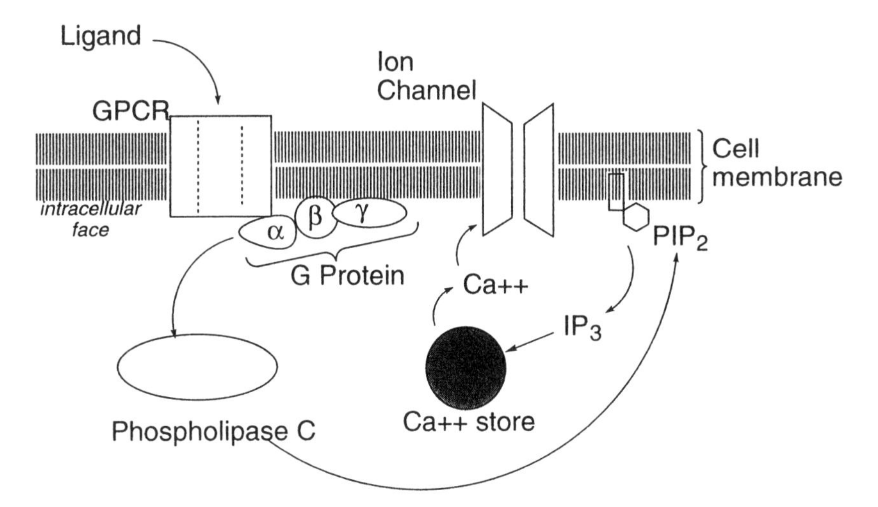

Figure 5. G Protein coupled receptor system in which the second messenger is a phospholipase which hydrolyses phosphatidylinositol to inositol trisphosphate, with subsequent release of intracellular calcium ions.

increase intracellular free calcium ion concentrations (*31*), and denatonium was shown to induce phospholipase C mediated release of inositol trisphosphate (*32*). Striem et al. found that sucrose and several other sweeteners stimulate the activity of adenylyl cyclase in tissue from rat tongue epithelium (*33,34*). The resulting cyclic AMP may exert its effect through blockade of potassium ion channels (*35*). There is also some evidence for involvement of sodium ion channels (*36*). Thus, much evidence points to the involvement of GPCR in sweet and bitter taste transduction. So far, however, no one has demonstrated a complete GPCR/G protein/second messenger pathway for any taste system.

The Proposed Rationale

How can we make sense of the various pieces of this puzzle? The structure-taste relationships, the taste interactions, the glimpses into G protein coupled receptor mechanisms all provide tantalizing clues, but do not yet constitute a clear picture. We describe a scenario based on available information and on analogy to other, more fully characterized G protein coupled receptor systems.

We propose that sweet and bitter tastes are detected by *families* of GPCR which are evolutionarily related and therefore similar to one another in some respects. There is ample precedent for multiple receptor types among G protein coupled receptors for neurotransmitters. Norepinephrine, epinephrine, dopamine and serotonin are detected by families of related receptors. For instance, adrenergic receptors (those activated by norepinephrine and epinephrine) were first classified as either α or β type, on the basis of selective activation or inhibition by specific drugs. As more drugs were discovered with other selectivities, these receptor types were later subdivided into $\alpha 1$, $\alpha 2$, $\beta 1$, and $\beta 2$ types. Sequencing of the genes for these receptors has recently shown that there exists a whole family of related, but not identical, receptors with varying responses to norepinephrine and related compounds (*37*). Some agonists and antagonists are non-selective, binding to all types of adrenergic receptors; other drugs are more selective, binding only to β receptors or only to $\alpha 2$ receptors, for example. Many antipsychotic drugs exhibit very broad activity, with affinity for adrenergic, dopaminergic, serotonergic, and muscarinic receptors.

It is thus reasonable to consider that sweet and bitter tastes are detected by families of receptor proteins which are evolutionarily related. Perception of sweetness and bitterness could be determined at the receptor level, at the G protein level, at the second messenger level, and at the ion channel level. At each of these levels, there could be sufficient homology and protein similarity to permit crossover between sweet and bitter modalities--a sweet compound might bind with lower affinity to a bitter receptor, inducing a conformation which activates it (causing bitter taste) or inducing a conformation which is inactive (suppressing bitter taste). Some sweeteners could act upon many or most sweet receptor types, while others act only upon subsets, leading to variation in taste quality. Such families of receptors could readily account for the phenomena of synergy and mixture suppression, and they could account for the tremendous diversity of sweet and bitter tastant structures. To fully sort out the various receptor types, we would have to have to have either some *selective antagonists* or gene/protein sequences of the different receptors.

With a multi-step mechanism, it is likely that different taste-active compounds could act at different stages of the transduction process. Some compounds may activate receptors, others could activate or block ion channels, and still others could influence second messenger events. Also, compounds acting upon different steps in the pathway could easily exhibit different temporal profiles. We expect that most known sweetness inhibitors must act at a *late* step in the transduction pathway, since they show no selectivity for classes of sweeteners but block essentially all sweet tasting compounds. On the other hand, numerous methods for blocking bitter taste have been described, and most of these are fairly specific for particular applications (see review by Roy, ref. 38).

A nagging question has been whether sugars act via receptors. In addition to their low potencies, sugars do not show stereoselectivity in taste; D-sucrose and L-sucrose are equally sweet (*39*). This, too, is unusual if a protein binding site is involved. It appears more likely to us that sugars act *indirectly* on the receptor system. At 0.1 M concentration, sugar is at a high enough concentration to affect water activity around the polar head groups of lipid membranes. If sugars alter the conformation or fluidity of taste cell membranes, they could indirectly change the conformations of receptor proteins, G proteins, adenylyl cyclase, phosphatidyl inositol, or ion channel proteins embedded in those membranes. Sweetness receptors may be GPCR which have been adapted by evolution to respond only to *high* concentrations of sugars (10^{-9} molar sugar solutions would not provide significant amounts of energy!). We can find precedent for activation via membrane effects in the case of mechanoreceptors: stretch-sensitive ion channels involved in the sense of touch respond to changes in *membrane tension* by opening or closing (*40*). If sweet receptors are GPCR which have evolved to detect sugar-induced changes in membrane properties, it must simply be coincidental that some small organic compounds can bind to these receptors and induce an active receptor conformation. It should not be surprising, then, that sugars are better able to induce maximal sweetness response, even though high potency sweeteners are active at lower concentrations.

The Implications

If the mechanism outlined above is operating, how can we use what we know to better control sweet taste, bitter taste, and sweet-bitter interactions?

Tasteless compounds. If we encounter tasteless compounds which are analogs of sweet or bitter compounds, we should test them as antagonists of sweet and bitter tastes. This was done routinely when the authors were carrying out sweetener research at The NutraSweet Company; several sweet and bitter taste inhibitors were discovered in this way. Such a procedure also led to the discovery of the sweetness inhibitor lactisole (*41*).

Masking bitterness. There are potentially *several* different ways to mask bitter taste:

- Experiment with a variety of different sweeteners or blends of sweeteners—there may be one or more with affinity for the relevant bitter receptors. High intensity sweeteners (e.g., sugars) and high potency sweeteners (e.g., aspartame) may both provide benefits.
- Use sweeteners with longer temporal profiles to more successfully mask lingering bitter tastes.
- Utilize specific bitter taste inhibitors (see reference 38).
- Add less intensely bitter compounds--if they act at the same receptor site and are *partial agonists*, they may produce submaximal bitterness.
- Increase viscosity, to decrease diffusion of compounds which act intracellularly or at ion channels beyond the tight junctions between taste cells.

Sweetness enhancement. To enhance sweet taste, use mixtures of two or more sweeteners. Aspartame plus acesulfame-K produces synergy, presumably because the two sweeteners act at different receptors or at different steps in the transduction pathway. Acesulfame-K plus saccharin does not produce synergy, probably because the two structurally related compounds act at the same site.

Blocking bitterness. If we can affect late steps in the transduction path, we are more likely to successfully block bitter taste. The step in the pathway which is affected may also be important in determining the temporal profile; this is particularly important for sweeteners, where deviation from the sucrose-like temporal pattern reduces consumer acceptance.

A major area for further sensory research is in the area of bitter taste (as unappealing as that may be!). As we continue to learn more about the ways in which sweet and bitter taste are perceived, we will expand our set of tools for controlling interactions between sweet and bitter taste.

Literature Cited

1. Tinti, J.-M.; Nofre, C. In *Sweeteners: Discovery, Molecular Design, and Chemoreception*, Walters, D. E., Orthoefer, F. T., and DuBois, G. E., Eds., American Chemical Society: Washington, 1991, 88-99.
2. DuBois, G. E.; Walters, D. E.; Schiffman, S. S.; Warwick, Z. S.; Booth, B. J.; Pecore, S. D.; Gibes, K.; Carr, B. T.; Brands, L. M. In *Sweeteners: Discovery, Molecular Design, and Chemoreception*, Walters, D. E.; Orthoefer, F. T.; DuBois, G. E., Eds., American Chemical Society: Washington, 1991, 261-276.
3. Payne, H. A. S. *Chem. Ind.* (London) **1988**, *22*, 721-723.
4. Whitney, G.; Maggio, J. C.; Harder, D. B. *Chem. Senses* **1990**, *15*, 243-252.
5. Jenner, M. R. In *Sweeteners: Discovery, Molecular Design, and Chemoreception*, Walters, D. E.; Orthoefer, F. T.; DuBois, G.E., Eds., American Chemical Society: Washington, 1991, pp 68-87.
6. Mazur, R. H.; Schlatter, J. M.; Goldkamp, A. H. *J. Amer. Chem. Soc.* **1969**, *91*, 2684-2691.

7. Glowaky, R. C.; Hendrick, M. E.; Smiles, R. E.; Torres, A. In *Sweeteners: Discovery, Molecular Design, and Chemoreception*, Walters, D. E.; Orthoefer, F. T.; DuBois, G. E., Eds., American Chemical Society: Washington, 1991, pp 57-67.
8. Audrieth, L. F.; Sveda, M. *J. Org. Chem.* **1944**, *9*, 89-101.
9. Fahlberg, C.; Remsen, I. *Chem. Ber.* **1879**, *12*, 469-473.
10. Clauss, K.; Jensen, H. *Angew. Chem. Internatl. Ed. Engl.* **1973**, *12*, 869-876.
11. Berlinerblau, J. *J. Prakt. Chem.* **1884**, *30*, 103-105.
12. Petersen, S.; Müller, E. *Chem. Ber.* **1948**, *81*, 31-38.
13. Muller, G. W.; Madigan, D. L.; Culberson, J. C.; Walters, D. E.; Carter, J.S.; Klade, C. A.; DuBois, G. E.; Kellogg, M. S. In *Sweeteners: Discovery, Molecular Design and Chemoreception*, Walters, D. E.; Orthoefer, F. T.; DuBois, G. E., Eds., American Chemical Society: Washington, 1991, pp 113-125.
14. Van der Wel, H.; Loeve, K. *Eur. J. Biochem.* **1972**, *31*, 221-225.
15. Morris, J. A.; Martenson, R.; Deibler, G.; Cagan, R. H. *J. Biol. Chem.* **1973**, *248*, 534-539.
16. Furukawa, S.; Tomizawa, Z. *J. Chem. Ind. Tokyo* **1920**, *23*, 342; *Chem. Abstr.* **1920**, 14, 2839.
17. Compadre, C. M.; Pezzuto, J. M.; Kinghorn, A. D.; Kamath, S. K. *Science* **1985**, *227*, 417-419.
18. Horowitz, R. M.; Gentili, B. *J. Agric. Food Chem.* **1969**, *17*, 696-700.
19. Acton, E. M.; Leaffer, M. A.; Oliver, S. M.; Stone, H. *J. Agric. Food Chem.* **1970**, *18*, 1061-1068.
20. Carr, B. T.; Pecore, S. D.; Gibes, K.M.; DuBois, G.E. In *Flavor Measurement*, Ho, C.-T.; Manley, C. H., Eds., Marcel Dekker, New York, 1993, pp 219-237.
21. Cohn, G. *Die Organischen Geschmacksstoffe*. Siemenroth, Berlin, 1914.
22. Warren, R. P.; Lewis, R. C. *Nature* **1970**, *227*, 77-78.
23. Clauss, K. *Liebigs Ann. Chem.* **1980**, 494-502.
24. Acton, E. M.; Leaffer, M. A.; Oliver, S. M.; Stone, H. *J. Agric. Food Chem.* **1970**, *18*, 1061-1068.
25. Muller, G. W.; Culberson, J. C.; Roy, G.; Ziegler, J.; Walters, D. E.; Kellogg, M. S.; Schiffman, S. S.; Warwick, Z. S. *J. Med. Chem.* **1992**, *35*, 1747-1751.
26. McBurney, D. H.; Bartoshuk, L. M. *Physiol. Behav.* **1973**, *10*, 1101-1106.
27. Akabas, M. H. *Int. Rev. Neurobiol.* **1990**, *32*, 241-279.
28. Dietzel, C.; Kurjan, J. *Cell* **1987**, *50*, 1001-1010.
29. Ross, E. M. *Neuron* **1989**, *3*, 141-152.
30. McLaughlin, S. K.; McKinnon, P. J.; Margolskee, R. F. *Nature* **1992**, *357*, 563-569.
31. Akabas, M. H.; Dodd, J.; Al-Awqati, Q. *Science* **1988**, *242*, 1047-1050.
32. Hwang, P. M.; Verma, A.; Bredt, D. S.; Snyder, S.H. *Proc. Natl. Acad. Sci. USA* **1990**, *87*, 7395-7399.
33. Striem, B. J.; Pace, U.; Zehavi, U.; Naim, M.; Lancet, D. *Biochem. J.* **1989**, *260*, 121-126.

34. Striem, B. J.; Yamamoto, T.; Naim, M.; Lancet, D.; Jakinovich, Jr., W.; Zehavi, U. *Chem. Senses* **1990**, *15*, 529-536.
35. Kinnamon, S. C. In *Sensory Transduction*, Corey, D. P.; Roper, S. D., Eds., Rockefeller University Press, 1992, pp 261-270.
36. Schiffman, S. S.; Lockhead, E.; Maes, F. W. *Proc. Natl. Acad. Sci. USA* **1983**, *80*, 6136-6140.
37. O'Dowd, B. F.; Lefkowitz, R. J.; Caron, M. G. *Ann. Rev. Neurosci.* **1989**, *12*, 67-83.
38. Roy, G. M. *Crit. Rev. Food Sci. Nutr.* **1990**, *29*, 59-71.
39. Szarek, W. A.; Jones, J. K. N. U.S. Patent 4,207,413, 1980.
40. French, A. S. *Ann. Rev. Physiol.* **1992**, *54*, 135-152.
41. Lindley, M. G. In *Sweeteners: Discovery, Molecular Design and Chemoreception*, Walters, D. E.; Orthoefer, F. T.; DuBois, G. E., Eds., American Chemical Society: Washington, 1991, pp 251-260.

Chapter 13

The Loss of Aspartame During the Storage of Chewing Gum

J.-P. Schirle-Keller[1], G. A. Reineccius[1], and L. C. Hatchwell[2]

[1]Department of Food Science and Nutrition, University of Minnesota, 1334 Eckles Avenue, St. Paul, MN 55108
[2]NutraSweet Kelco Company, 601 East Kensington Road, Mt. Prospect, IL 60056–1300

The loss of aspartame was determined in several different model chewing gum systems during storage. The model gum systems were composed of selected chewing gum components (gum base, sweeteners, aspartame and flavor). The flavor included *t*-2-hexenal, carvone, menthol, benzaldehyde, cinnamic aldehyde and β-ionone in equal proportions. Aspartame was found to be lost the most rapidly from model systems containing flavoring. For example, all aspartame was lost from an flavor/aspartame model system in as little as five days (35 °C). Aspartame losses were observed in model systems without flavoring but at much reduced rates.

There have been numerous studies on the stability of aspartame (APM) in foods (e.g., *1-6*). APM has generally been found to be most stable at low a_w, low temperatures of storage and moderate pH (most stable at pH 5). There are two mechanisms of APM loss – one involving the hydrolysis of the methyl ester with subsequent cyclization to the diketopiperazine, and the other involving the reaction of its amine group with carbonyl compounds. While the majority of research on APM stability has focused on its decomposition to diketopiperazine (*4,7*), it is readily lost via carbonyl/amine reactions.

There is little information in the literature on flavor/APM interactions other than some work on vanillin (potential dairy applications) and diet colas. Cha et al. (*8*) reported that vanillin will readily react with APM depending on temperature and reaction conditions. Higher temperatures and reduced water in the model system (methanol/water model systems) increased the reaction rate. Berte et al. (*9*) observed major losses of vanillin and cinnamic aldehyde (68 and 75%, respectively) in APM-sweetened diet colas following only two weeks of storage. APM losses in these same beverages were approximately 21% over four months of storage.

0097–6156/96/0633–0143$15.00/0

Despite these losses, Homler (*1*) noted that APM-sweetened carbonated beverages remained acceptably sweet over six months storage.

In addition to the scientific literature and more directly relevant to this current study, there is information in the patent literature on APM stability in chewing gums (*10-12*). Cea et al. (*10*) presented data on the stability of encapsulated APM in chewing gums. The APM was encapsulated via spray drying (gum acacia or modified starch carriers) or fluidized bed coating (in a hydrocolloid such as gelatin, modified cellulose or polyvinylpyrrolidone). Losses of encapsulated APM ranged from 10 to 60% when a cinnamon flavored gum was stored 18 weeks at 37 °C. No data were presented on APM which was not encapsulated. The patent of Cherukuri et al. (*11*) focused on APM reactions which lead to color problems in chewing gum. No quantitative data were presented on APM losses. Greenberg and Johnson (*12*) chose to protect the flavor and APM by using acetals of the aldehydic flavor components instead of the free aldehydes themselves. Assuming there is no hydrolysis of the acetals in the gum during storage, there would be no reaction between these acetals and APM, thereby reducing losses of both. They presented data showing virtually complete losses of APM (from 0.44% to 0.02%) over a period of 21 days storage (30 °C) when cinnamic aldehyde was used as the flavoring. Losses of APM when cinnamaldehyde propylene glycol acetal was substituted for cinnamic aldehyde were less that 10% over a comparable storage period.

Reactions between flavor and APM in gum would lead the loss of sweetness and/or flavor. Therefore, it was the goal of this study to determine what flavors will react with APM and determine the extent of their reaction. The model systems reported in this chapter varied from simple flavor/APM mixes to complete gum systems. Thus the contribution of flavor, sweetener and gum base to APM losses could be determined.

Materials and Methods

Gum Materials. The following ingredients for chewing gum preparation were obtained from NSC Technologies: 42/43 Corn Syrup (ADM Corn Sweeteners, Decatur, IL), glycerin (Humko Chemical Division, Memphis, TN), White Satin Powdered Sugar (Chicago Sweeteners, Hillside, IL), Lycasin (R) 85%. (Chicago Sweeteners, Hillside, IL), Sorbitol FCC Neosorb 60W (Roquette Corp., Gurnee, IL), Nova gum base (L. A. Dreyfus Co., South Plainfield, NJ) and Aspartame (APM, NSC Technologies, Mt. Prospect, IL.). These ingredients were used to formulate eight model systems for study (Table 1).

Flavor System. The model flavor solution contained equal proportions of *t*-2-hexenal (green-grassy character), benzaldehyde (cherry), menthol (mint), *l*-carvone (spearmint), cinnamaldehyde (cinnamon) and β-ionone (raspberry). All compounds were obtained from Aldrich Chemical Co. (Milwaukee, WI). These compounds were mixed together and stored in a refrigerator at 4 °C until used in the model gum systems.

Preparation of Chewing Gum Model Systems. An outline of the parameters for the eight chewing gum model systems used in this study is presented in Table I. The complete chewing gum model systems (systems 1 and 2) were prepared by first weighing the gum base in a 125-mL beaker and heating it in a microwave oven 5 min on high to melt it. The beaker of melted gum base was put on a scale and the sweetener solution was directly weighed into the molten gum. The sweetener solution had the following composition: sorbitol, 268.45 g (53. 69%); Lycasin, 90.60 g (18.12%); glycerin, 35.25 g (7.05%) and corn syrup, 105.70 g (21.14%). The gum:sweetener mixture was stirred by hand until homogenous. APM was weighed and mixed with half of the desired glucose. This glucose/APM blend was mixed into the gum base when the gum had cooled to 42 °C. The model flavor system, when included in the formulation, was added when the gum system had cooled to 37 °C. The gum system was mixed well and then the last half of the glucose was added and thoroughly mixed. Powdered sugar (5 g) was used to cover a cutting board and the molten gum was poured on top of it. The gum was allowed to cool and dry slightly. An additional 3 g of powdered sugar was added to the gum system to facilitate handling of the gum. The gum was then cut into pieces and stored in 200-mL mason jars.

Model systems 3 and 4 consisted of the sweetener system (sweetener solution, glucose and APM) with and without flavor. These systems were prepared by first weighing the sweetener solution into a 125-mL beaker, warming it in a microwave oven and mixing well to ensure homogeneity. One half of the desired amount of glucose and all the APM (preblended) were blended into this solution when it had cooled to 42 °C. The flavor solution, when included (system 3), was added when the sweetener system had cooled to 37 °C. The system was stirred well and the remaining glucose was blended into the mix. To be consistent, 8 g of powdered

Table I. Model System Formulations

	Model System Composition (%)					
System	*Sweetener*	*Sucrose*	*Gum Base*	*Flavor*	*APM*	*Total*
1	42.73	28.97	25.00	3.00	0.30	100
2	44.06	29.86	25.77	—	0.31	100
3	56.98	38.62	—	4.00	0.40	100
4	59.35	40.23	—	—	0.42	100
5	—	—	88.34	10.60	1.06	100
6	—	—	98.81	—	1.19	100
7	—	—	—	90.91	9.09	100
8	—	—	—	—	100	100

sugar was finally mixed into the model system. This molten mass was placed in a 200-mL mason jar for storage.

Model systems 5 and 6 contained gum base and APM with and without flavor (no sweetener, glucose or powdered sugar). These model systems were prepared similarly to the complete gum system, except that all sweeteners but APM were left out of the formulation. The APM was added directly to the molten gum (at 42 °C). Flavor, when included (system 5), was added at 37 °C. The molten sticky mass was blended well and then transferred to a 200-mL mason jar.

The system containing flavor and APM (system 7) was made by simply mixing the ingredients together at room temperature. This system was not heated. All model systems were made in duplicate. One mason jar of each system was stored at 20 °C while the other was stored at 35 °C. A sample (2.5 g) of each system was taken prior to storage to represent time 0.

Sampling Periods. Four samples (ca. 2.5 g) of each system at each temperature were taken at 2, 4 and 8 days and 2, 4, 7 and 16 weeks storage (stored at -20 °C until extraction). For systems 7 and 8, 0.5-g and 0.05-g samples were taken, respectively, at each sampling period.

Analytical Methods

APM Extraction. Samples (2.5 g) were frozen in dry ice prior to extraction. The frozen sample was mixed with 5 g Celite and then finely ground with dry ice in a mortar and pestle. The mix was poured into a pre-weighed 50-mL Nalgene centrifuge tube where the dry ice was allowed to sublime prior to weighing. Ten mL of hexane was added to the sample followed by 25 mL of extraction buffer. (The extraction buffer was prepared by dissolving 25 mg of sodium phosphate (monobasic) in 800 mL water (1 L volumetric flask), adjusting the pH to 2.5 with phosphoric acid and then making the solution to 1 L). The centrifuge tubes were placed on a mechanical shaker (Burrell model 75, Pittsburgh, PA) and were shook for 18 hrs using the maximum shaking angle (10^0) at room temperature. Tubes were then weighed and balanced with hexane so they could be centrifuged (Beckman model J2-21, Palo Alto, CA, 10 min at 2000 rpm). The supernatant (hexane) was removed using a water aspirator (with a trap in the line) and the water phase was withdrawn into a 5-mL hypodermic syringe and filtered through a 0.45μm filter fitted to the syringe. The extract was poured into a 4-mL HPLC automatic sampler vial, degassed, capped and immediately stored in the tray of the autosampler.

High Performance Liquid Chromatography. A Waters HPLC system (Waters and Associates, Milford, MA) was used with a 10% acetonitrile in phosphate buffer (pH = 2.5) mobile phase. The separation was isocratic and was performed on a Microbondapak C18 column (Waters and Associates, Milford, MA) protected by a GuardPak precolumn (Waters and Associates, Milford, MA). APM was detected and quantified at 220 nm. Separation parameters were: column flow, 2 mL/min; run time, 20 min; and injection volume, 20 μL. Before each sample set (given day of storage or temperature), an APM standard solution was injected in order to calibrate the instrument. At the end of each day, the column was rinsed with 100 mL of 10/90 methanol/water mobile phase, followed by 50 mL of pure methanol.

Results

APM Losses in Flavor Systems. Pure APM (no gum components or added flavor) was stable at both temperatures of storage over the study period (Figure 1). However, when the flavor system was added to the APM, APM was completely lost from the system after ca. 8 days at 20 °C or 5 days at 35 °C. Work reported in a related publication (*13*) discusses the interaction of flavor compounds with APM. It is noted in this paper that APM readily reacts with unsaturated aldehydes to yield Schiff's bases. Since this model system contains *t*-2-hexenal, cinnamic aldehyde and benzaldehyde, the loss of APM via reaction is anticipated.

Aspartame Loss In Gum Base System. The gum systems did not contain any sweetener, only gum base and APM with or without the flavor solution. It was particularly difficult to uniformly distribute the APM and flavor into this base because the gum base without the sweetener system tended to solidify very quickly. The problems in blending were evident, since the color of the system that developed during storage was not uniform throughout the sample. This is likely a primary factor responsible for the large variability found in the data. A second factor was that the gum changed in physical properties after the first two weeks of storage; it became drier in appearance and less elastic. This may have influenced the extraction efficiency of the APM method. Thus the trends in these data must be evaluated in the presence of substantial variability. This problem was even worse when there was no flavor to plasticize the gum mass. Although the data are extremely variable, APM appears to be relatively stable in the gum system without flavor (Figure 2).

In the gum base system containing flavor (system 5), there was again a rapid degradation of APM. At 20 °C, the APM concentration dropped by ca. 80% over the 2 first weeks of the experiment and after 4 months, APM losses approached 92% of the initial amount. At 35 °C, APM was totally degraded after only 10 days. APM was lost from the gum base systems more slowly than from the simple APM and flavor mixtures.

APM Losses in Sweetener Systems. There was no apparent loss of APM from the sweetener system (no gum base) at 20 °C but ca. 15% loss at 35 °C (Figure 3) over the duration of the study. It is of interest that there was a small loss of APM at 35 °C even though there was no flavor in this system (contrary to what was observed for the pure APM system).

APM losses increased greatly when flavor was included in the model system. More than 90% of the APM was lost over the storage period when flavor was included, irrespective of storage temperature. Even though total losses from the sweetener/flavor/APM system were about the same as in the flavor/APM system, losses were more gradual from the sweetener-containing system, suggesting that the sweetener was adding some barrier to reaction.

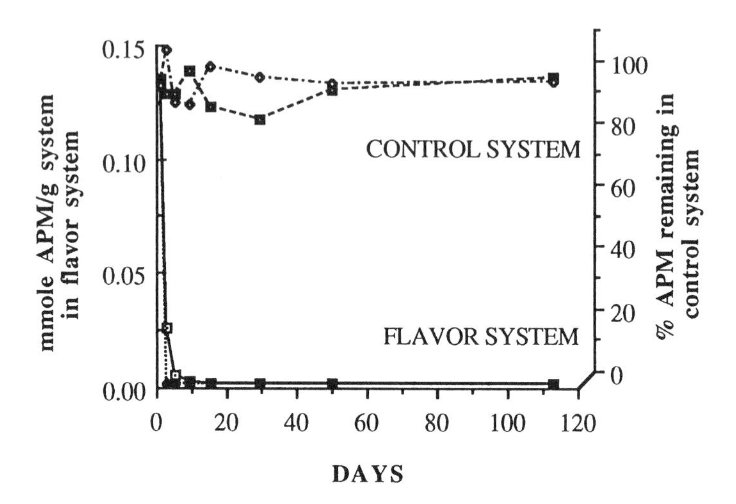

Figure 1. Loss of aspartame during the storage of aspartame or an aspartame/flavor mixture (20° and 35 °C).

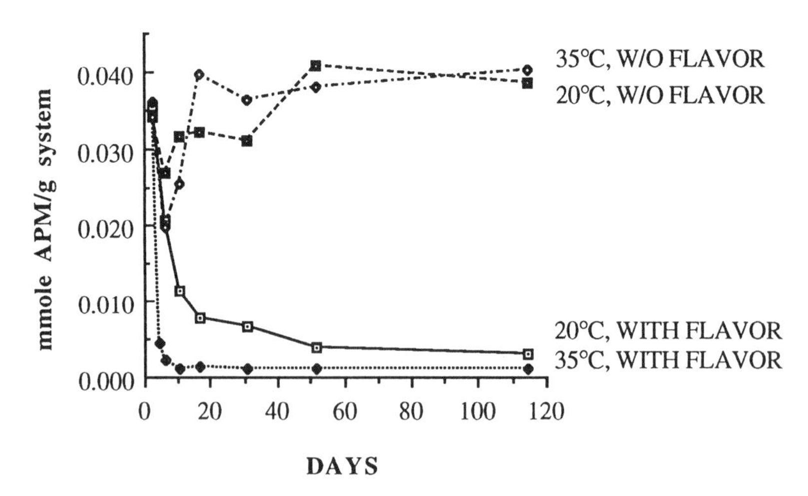

Figure 2. Loss of aspartame in a gum base during storage with and without flavor (20° and 35 °C).

Aspartame Loss In Complete Chewing Gum Systems. This complete system contained the sweetener blend (corn syrup, Lycasin, sorbitol and glycerin), glucose, gum base and APM with or without the flavor solution. In system 2 (containing no flavor), APM losses were ca. 15% over the storage study (20 °C). At 35 °C storage, APM losses were ca. 70% over this period (Figure 4). APM losses appeared to be greatly enhanced in the complete gum system (which contains the sweetener system) versus the simpler gum base system.

In system 1, containing flavor, APM degraded quickly at both storage temperatures. APM losses approached 25% and 95% at 20 °C and 35 °C over the 2 first weeks of the study, respectively. APM was totally degraded after 7 weeks of storage at 35 °C but losses leveled off at ca. 78%. for the samples stored at 20 °C.

Discussion

APM losses were consistently greater from the systems which contained added flavoring. As was noted earlier, we have found APM to readily react with the unsaturated aldehydes included in our flavoring mixture (*13*) and thus this observation is expected. The fastest loss rates were observed for the simple mixture of flavoring and APM. Loss rates decreased when other ingredients were added to the system. This may have occurred due to the dilution effects of dispersing the flavor and APM throughout a model system or perhaps reduced diffusion rates as the model gum systems contributed some barrier to diffusion. An additional factor is the role of various gum ingredients on the water activity (a_w) of the system. If APM losses are due primarily to the reaction of our flavor system with the amine function of the APM via Shiffs base formation, we would expect a_w to play a significant role in determining this reaction rate. Since we did not measure the a_w in any of our sytems, we can not determine the influence of a_w on APM loss.

APM may well have been lost by mechanisms which are independent of the presence or absence of flavor. This is evident since we noted differences in the stability of APM in the model systems which did not contain any flavoring. These losses may have been due to a reaction of the APM with glucose (Maillard reaction), or its spontaneous decomposition by loss of the methyl ester and then hydrolysis to individual amino acids, or cyclization to 3-carboxymethyl-6-benzyl-2,5-dioxo-piperazine (diketopiperazine) (*7, 14*). APM is less stable at higher pH, a_w and/or temperature (*1, 4, 6*).

We did not conduct any experiments to determine the relative importance of the various degradation pathways for APM. Thus we can not comment on the major pathway for APM losses other than note that the presence of flavorings containing unsaturated aldehydes results in major losses of APM in gum and model systems containing gum components. Clearly, flavorings containing unsaturated aldehydes will react with APM resulting in the rapid loss of APM. Additionally, we can not comment on the relative importance of diffusion, dilution or water activity in determining the reaction rates of APM and flavoring (or reducing sugars) or its spontaneous degradation.

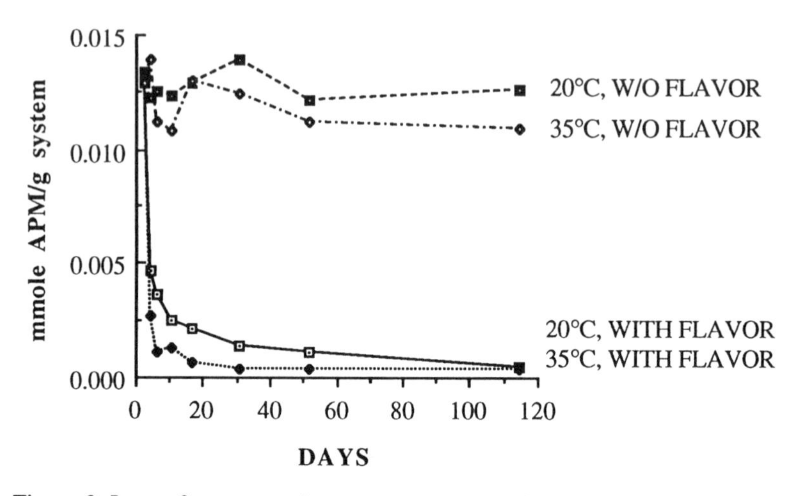

Figure 3. Loss of aspartame in a gum sweetener mixture during storage with and without flavor (20° and 35 °C).

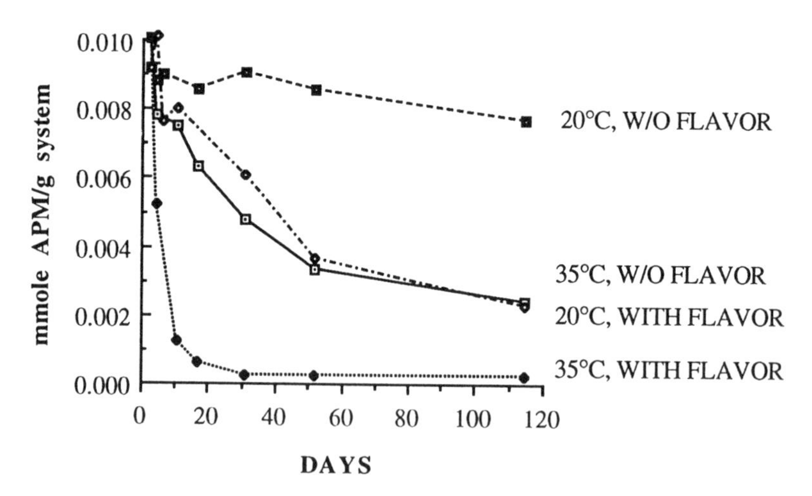

Figure 4. Loss of aspartame in a complete gum formulation during storage with and without flavor (20° and 35 °C).

Conclusions

APM stability during storage of all chewing gum model systems was found to be greatly reduced by the presence of the flavor model system. Other work in our laboratory has shown that the APM is stable in the presence of most flavor compounds except unsaturated aldehydes (*13*).

Some losses of APM were noted in gum model systems which did not contain any flavor compounds. The mechanism(s) of APM loss in these systems is less certain. However, it is well documented in the literature that APM will spontaneously degrade with time. The rate of degradation is related to a_w, storage temperature and pH.

Literature Cited

1. Homler, B. E. *Food Technol.* **1984**, *38*, 50.
2. Prudel, M.; Davidkova, E.; Davidek, J.; Kminek, M. *J. Food Sci.* **1986**, *51*, 1393.
3. Redlinger, P. A.; Setser, C. S. *J. Food Sci.*, **1987**, 52, 1391.
4. Stamp, J. A. *Kinetics and Analysis of Aspartame Decomposition Mechanisms in Aqueous Solutions using Multiresponse Methods;* Ph.D. thesis, Dept. of Food Science and Nutrition, Univ. of Minnesota, St. Paul, MN, 1990.
5. Prudel, M.; Davidkova, E. *Die Naturung,* **1985**, *29*, 381.
6. Bell, L. N.; Labuza, T. P. *J. Food Sci.*, **1991**, *56*, 17.
7. Furda, I.; Malizia, P. D.; Kolor, M. B.; Vernier, P. J. *J. Agric. Food Chem.*, **1975**, *23*, 340.
8. Cha, A. S.; Ho, C.-T.; Huang, T. C.; Sotiros, N. In *Frontiers of Flavors;* G. Charalambous, Ed.; Elsevier Publ.: Amsterdam, 1988, pp. 233-244.
9. Berte, F.; Tateo, F.; Triangeli, E.; Panna, E.; Verderio, E. In *Frontiers of Flavors;* G. Charalambous, Ed.; Elsevier Publ.: Amsterdam, 1988, p. 217-231.
10. Cea, T.; Posta, J. D.; Glass, M. *Encapsulated APM and method of preparation*, U.S. Patent 4 384 004, 1983.
11. Cherukuri, S. R.; Mansukkami, G.; Kapakkamannil, C. J. *Stable cinnamon-flavored chewing gum composition*, U.S. Patent 4 722 845, 1988
12. Greenberg, M. J.; Johnson, S. D. *Method of stabilizing peptide sweeteners in cinnamon-flavored chewing gums and confections*, U.S. Patent 5 167 972, 1992.
13. Schirle-Keller, J. P.; Reineccius, G. A.; Hatchwell, L. C. *J. Agric. Food Chem.,* **1996**, (submitted).
14. Stamp, J.A.; Labuza, T.P. *J. Food Sci.,* **1989**, *54*, 1043.

Chapter 14

Sorption and Diffusion of Flavors in Plastic Packaging

Phillip T. DeLassus

Dow Chemical Company, 438 Building, Midland, MI 48667

Flavors can be lost from an originating food to a plastic package by the physical process of dissolving. The thermodynamic character of this event is described by the equilibrium partition (sorption) of the flavor between food and package. The kinetic character of this event is described by the penetration into the package wall (diffusion). Together, the sorption and diffusion can model the interaction of flavors with a plastic package. Example data will be given and discussed. The important role of the glass transition temperature in the plastic will be noted.

This chapter will review those interactions between food and plastic packaging that result from sorption and diffusion of flavors into the polymer. The sorption and diffusion events are component parts of the greater phenomenon of permeation. This discussion will begin with permeation and move to sorption and diffusion. For simple cases, Fick's First Law represents an adequate model for permeation. Equation 1 gives Fick's First Law in terms that are familiar to packaging professionals:

$$\frac{\Delta M_x}{\Delta t} = \frac{PA\Delta p_x}{L} \quad (1)$$

where $\Delta M_X/\Delta t$ is the rate at which permeant x passes through a polymer film of area A and thickness L when a difference in the partial pressure of the permeant, Δp_X, exists from one side of the film to the other. The permeability coefficient (or, more commonly, simply the "permeability") P completes the equation. This equation is useful for steady-state permeation after "break through" has occurred. For small molecules such as oxygen or water in most polymer films, the time to reach steady state is small compared to the total time in the package. Furthermore, the amount of

0097–6156/96/0633–0152$15.00/0

sorption of the permeant in the package wall is either irrelevant (oxygen from the room) or very small. This explains why Fick's First Law has had such an important role in describing barrier packaging for so many situations.

Equation 2 defines the permeability in terms of more fundamental properties – the diffusion coefficient, D, and the solubility coefficient, S.

$$P = D \cdot S \tag{2}$$

The diffusion coefficient is a kinetic term that describes how fast permeant molecules move in the polymer host. The diffusion coefficient is a function of the tightness of the polymer which is more rigorously related to the free volume, the size of the permeant, the temperature, and the amount of crystallinity. Polymers with small free volumes have low diffusion coefficients. Large or bulky permeants have low diffusion coefficients. The diffusion coefficient rises exponentially with rising temperature. Crystallinity in a polymer leads to lower diffusion coefficients; however, this effect is highly overrated. The diffusion coefficient can be affected by the amount of permeant that has dissolved into the polymer if the concentration is high. Typically, large concentrations of a permeant can plastisize the polymer and increase the diffusion coefficient. When dealing with a specific polymer, the free volume and the level of crystallinity are not likely to vary much from one application to another. Additionally, the activities of most flavor compounds in foods are low; hence, the equilibrium concentration in a package is likely to be very low. This means that the principle external variable for a given permeant is the temperature.

The solubility coefficient is a thermodynamic term that describes how much of a permeant will dissolve into a polymer at equilibrium. The solubility coefficient is a function of the interactions between the polymer and the permeant, the activity of the flavor compound, and the temperature. The simple rule "like dissolves like" is a good place to start for dissolving flavors into a polymer. If the polymer and the permeant have similar solubility parameters, the solubility coefficient will tend to be high. Typically as the activity of the permeant increases to near an activity of unity, the solubility coefficient rises. However, at activities less than about 0.25, Henry's Law typically holds well, and the solubility coefficient does not vary (*1*). Depending on the heat of solution, the solubility coefficient can rise or fall with increasing temperature. Typically, this effect is not as strong as the change of the diffusion coefficient with temperature. The solution occurs in the amorphous regions of the polymer.

The understanding of the permeation process is not yet complete enough to rely on theoretical calculations for the permeabilities, diffusion coefficients, and solubility coefficients for flavor compounds in polymers. Hence, experimental data are required. All three of the permeation parameters — P, D, and S — can be determined in a single experiment. Figure 1 is a representation of the experimental output of a permeation experiment where the instrumental response on the vertical axis represents the instantaneous rate of flux through a sample film. The diffusion coefficient can be calculated with equation 3 where $t_{1/2}$ is the time needed for the experimental response to rise half way to steady state after a clean film is exposed to a flavor compound.

$$D = \frac{L^2}{7.2t_{1/2}} \tag{3}$$

The permeability can be calculated using equation 1 with the steady-state flux rate that is shown by the constant response toward the right side of Figure 1 and the known experimental variables of film thickness and area plus the difference in flavor pressure from one side of the film to the other. The solubility coefficient can be calculated with equation 2.

Applications

With reliable values for *P*, *D*, and *S*, some simple calculations can be performed to estimate the interaction of a flavor from a food and a proposed package. With small permeants such as oxygen and water, the time needed to reach steady state is short, and the supply of permeant, e.g., oxygen from the atmosphere or water from the food, is large compared to any sorption which could occur in the package. Hence, simple calculations with equation 1 for the permeation rate are useful.

However, with larger permeants such as flavors, more information can be calculated. Since the diffusion coefficients are much smaller for large flavor molecules than for the smaller molecules, the time to reach steady state can be much longer. Equation 4 can be used to estimate the time to reach steady state, t_{SS}.

$$t_{ss} = \frac{L^2}{4D} \tag{4}$$

Here knowledge of the diffusion coefficient and the thickness of the package wall is needed before equation 1 may be used to calculate the steady-state permeation rate. Steady-state permeation may not be reached during the anticipated storage time for some combinations of flavors and films.

The amount of flavor compound that will be absorbed by the package wall can be estimated if the solubility coefficient and the partial pressure of the flavor in the food, p_X, are known. Equation 5 gives the average concentration in a package wall at steady state when the partial pressure of the flavor compound outside the package is negligible.

$$C_{avg} = 0.5\ S\ p_X \tag{5}$$

The amount of flavor sorbed by the package, ΔM_{sorb} will be C_{avg} times the volume of the package wall, *V*.

$$\Delta M_{sorb} = C_{avg} \cdot V \tag{6}$$

Because the solubility coefficients for flavors in polymers can be quite large, the quantity sorbed can represent an appreciable fraction of the available flavor. This phenomenon is called "scalping" and is especially important when some flavor components in a food are sorbed more than other components.

Before steady state is achieved, the amount of flavor sorbed will be less since a portion of the wall thickness will not have fully participated in the sorption. The effective penetration depth, L_{eff}, of the sorption of flavor can be estimated with equation 7 where t_{stor} is the time the food has been in storage in the plastic package.

$$L_{eff} = 2(Dt_{stor})^{1/2} \quad (7)$$

This effective thickness should be used to calculate the effective volume of the package wall for equation 6.

Example Data

Table I contains a list for comparing the values of diffusion coefficients of large and small molecules in a few common polymers. Comparisons can be made for high density polyethylene and for a vinylidene chloride copolymer. Here, the diffusion coefficients for flavor molecules are about 1000-times smaller than for small molecules. This is typical.

Table I. Diffusion Coefficients and Solubility Coefficients at 25 °C

Penetrant	Polymer	D *m^2/s*	S *kg/m^3 Pa*
Oxygen	polyethylene terephthalate	3.0×10^{-13}	9.8×10^{-7}
Oxygen	high density polyethylene	1.7×10^{-11}	6.6×10^{-7}
Oxygen	vinylidene chloride copolymer	1.5×10^{-14}	3.5×10^{-7}
CO_2	acrylonitrile copolymer	1.0×10^{-13}	1.6×10^{-6}
CO_2	poly(vinyl chloride)	8.9×10^{-13}	3.4×10^{-6}
CO_2	vinylidene chloride copolymer	1.4×10^{-14}	1.1×10^{-6}
d-Limonene	high density polyethylene	7.0×10^{-14}	0.3
d-Limonene	vinylidene chloride copolymer	3.0×10^{-18}	0.6
Methyl salicylate	Nylon 6	2.1×10^{-17}	0.9
Methyl salicylate	vinylidene chloride copolymer	5.8×10^{-16}	0.3

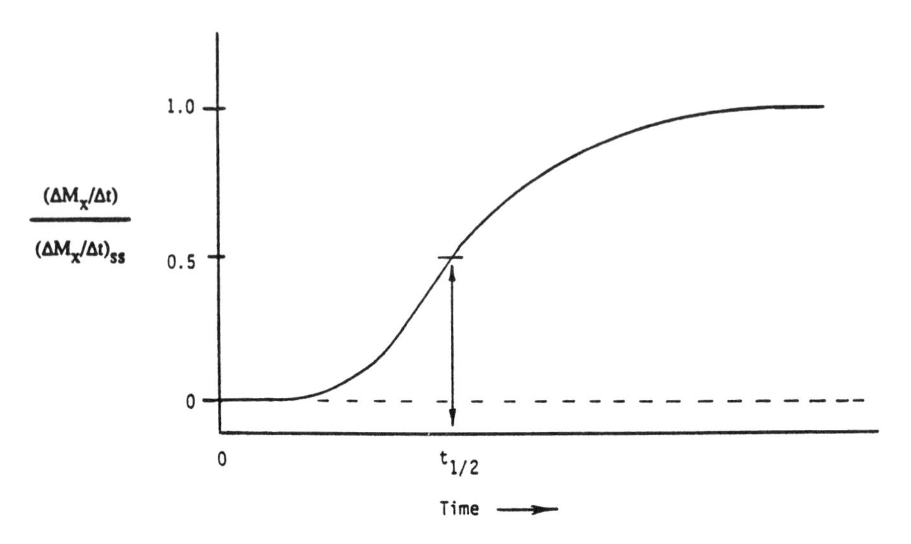

Figure 1. Relative transport rate as a function of time. (Reproduced with permission from ref. 7. Copyright 1990 American Chemical Society.)

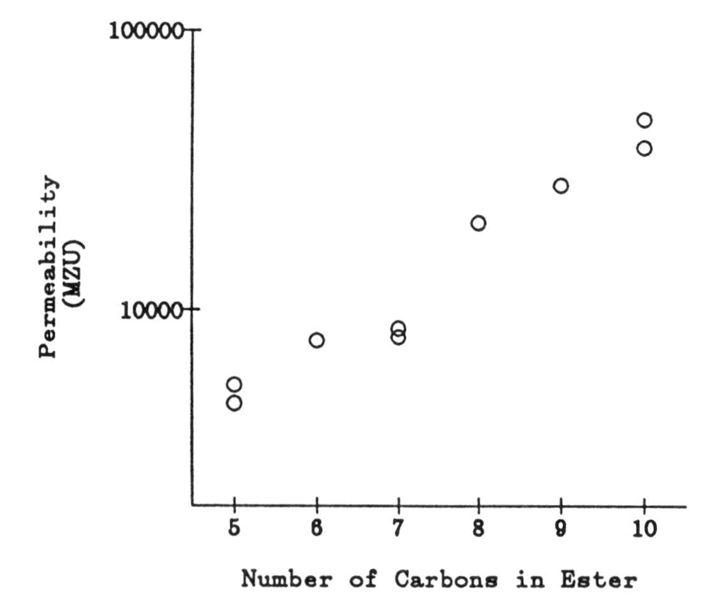

Figure 2. Permeabilities of linear esters at 85 °C in a vinylidene chloride copolymer film. (Reproduced with permission from ref. 7. Copyright 1990 American Chemical Society.)

Table I also contains data for comparing the values of solubility coefficients of large and small molecules in the same polymers. The solubility coefficients for flavor molecules are about 100,000-times larger than for small molecules. This is typical although great variations have been observed. Table II is a wider compilation of permeation data. While a great deal more data such as these are available, unfortunately the total is still much less than is needed for complete design work. Specific experiments can be done or trends can be observed and used to supply missing data.

A study with a family of linear esters is an example of using trends to supply missing data. A series of linear esters with from five to ten carbon atoms were studied in a few polymers. Figure 2 contains the permeability results for a vinylidene chloride copolymer at 85 °C. This high temperature was required since this is a barrier polymer with very low permeability at lower temperatures. The permeability increases with increasing size of ester. This is counterintuitive, but easily explained with the data in Figures 3 and 4. Figure 3 shows the diffusion coefficient for the same combinations. The diffusion coefficient decreases with increasing size of the ester. However, the solubility coefficient increases rapidly with increasing size of the ester. The change in the solubility coefficient is greater than the change in the diffusion coefficient; hence, the permeability increases with increasing ester size. Equivalent results have been found for other polymers and for other permeant families. A thorough review of this topic has not been written. However, the reader will find references 1, 9, 10, 11, and 12 helpful.

Glassy Polymers

The data given in the previous paragraphs are nearly all for permeation in rubbery polymers, i.e., polymers above their glass-transition temperatures, T_g. Data for flavor permeation in glassy polymers are sparse for a simple reason, namely, the diffusion coefficients are so low. Hence, permeabilities are frequently too low to measure and steady state is achieved only after a very long time.

Figure 5 presents diffusion data gathered by Berens (*9*). For rigid polyvinyl chloride (below T_g), the diffusion coefficients decrease sharply as the sizes of the permeants increase. For plasticized polyvinyl chloride (above T_g), the diffusion coefficients decrease more slowly as the sizes of the permeants increase. The mean diameter of hexane is about 0.6 nm. Most flavor molecules will be larger than hexane.

Figure 5 does not contain data for glassy polymers with permeants larger than 0.6 nm. The reason is that the time for experiments with a diffusion coefficient of $10^{-22} m^2/s$ in a hypothetical film only 1.0 μm (0.04 mil) thick would be 80 years. Smaller diffusion coefficients would take longer. Diffusion coefficients in rubbery polymers can be ten orders of magnitude larger, and the experimental times would be proportionately shorter.

Similar results have been found with other glassy polymers. Diffusion coefficients in polystyrene, polymethyl methacrylate, the polycarbonate of bisphenol-A, and polyethylene terephthalate have the same sharp dependency on

Table II. Examples of Permeation of Flavor and Aroma Compounds in Polymers at 25 °C[a], Dry (Refs. 6-8)

Flavor/aroma compound	P, *MZU*[b]	D, m^2/s	S, $kg/(m^3 \cdot Pa)$
Low-density polyethylene			
Ethyl hexanoate	4.1×10^{6}	5.2×10^{-13}	7.8×10^{-2}
Ethyl 2-methylbutyrate	4.9×10^{5}	2.4×10^{-13}	2.3×10^{-2}
Hexanol	9.7×10^{5}	4.6×10^{-13}	2.3×10^{-2}
trans-2-Hexenal	8.1×10^{5}		
d-Limonene	4.3×10^{6}		
3-Octanone	6.8×10^{6}	5.6×10^{-13}	1.2×10^{-1}
Propyl butyrate	1.5×10^{6}	5.0×10^{-13}	3.0×10^{-2}
Dipropyl disulfide	6.8×10^{6}	7.3×10^{-14}	9.3×10^{-1}
High-density polyethylene			
d-Limonene	3.5×10^{6}	1.7×10^{-13}	2.5×10^{-1}
Menthone	5.2×10^{6}	9.1×10^{-13}	4.7×10^{-1}
Methyl salicylate	1.1×10^{7}	8.7×10^{-14}	1.6
Polypropylene			
2-Butanone	8.5×10^{3}	2.1×10^{-15}	4.0×10^{-2}
Ethyl butyrate	9.5×10^{3}	1.8×10^{-15}	5.3×10^{-2}
Ethyl hexanoate	8.7×10^{4}	3.1×10^{-15}	2.8×10^{-1}
d-Limonene	1.6×10^{4}	7.4×10^{-16}	2.1×10^{-1}
Vinylidene chloride copolymer			
Ethyl hexanoate	570	8.0×10^{-18}	0.71
Ethyl 2-methylbutyrate	3.2	1.9×10^{-17}	1.7×10^{-3}
Hexanol	40	5.2×10^{-17}	7.7×10^{-3}
trans-2-Hexenal	240	1.8×10^{-17}	0.14
d-Limonene	32	3.3×10^{-17}	9.7×10^{-3}
3-Octanone	52	1.3×10^{-18}	0.40
Propyl butyrate	42	4.4×10^{-18}	9.4×10^{-2}
Dipropyl disulfide	270	2.6×10^{-18}	1.0
Ethylene-vinyl alcohol copolymer			
Ethyl hexanoate	0.41	3.2×10^{-18}	1.3×10^{-3}
Ethyl 2-methylbutyrate	0.30	6.7×10^{-18}	4.7×10^{-4}
Hexanol	1.2	2.6×10^{-17}	4.6×10^{-4}
trans-2-Hexenal	110	6.4×10^{-17}	1.8×10^{-2}
d-Limonene	0.5	1.1×10^{-17}	4.5×10^{-4}
3-Octanone	0.2	1.0×10^{-18}	2.0×10^{-3}
Propyl butyrate	1.2	2.7×10^{-17}	4.5×10^{-4}

[a]Values for vinylidene chloride copolymer and ethylene-vinyl alcohol are extrapolated from higher temperatures.
[b]MZU = 10^{-20} $kg \cdot m/(m^2 \cdot s \cdot Pa)$

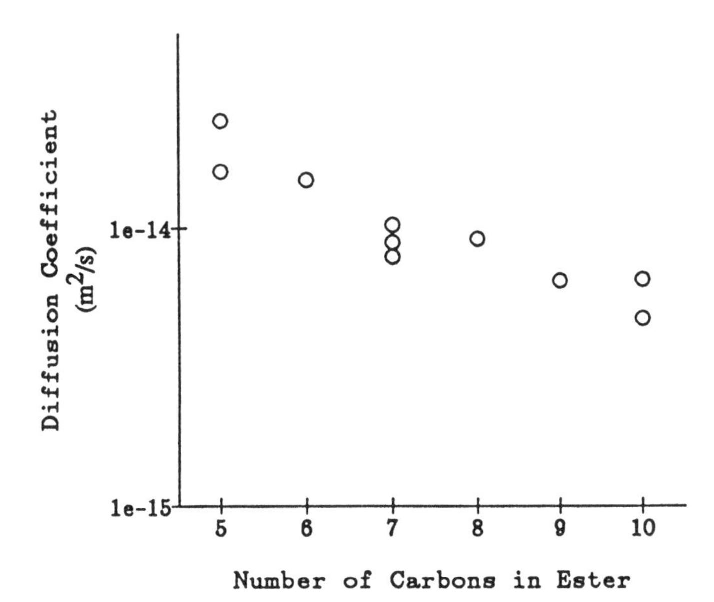

Figure 3. Diffusion coefficients of linear esters at 85 °C in a vinylidene chloride copolymer film. (Reproduced with permission from ref. 7. Copyright 1990 American Chemical Society.)

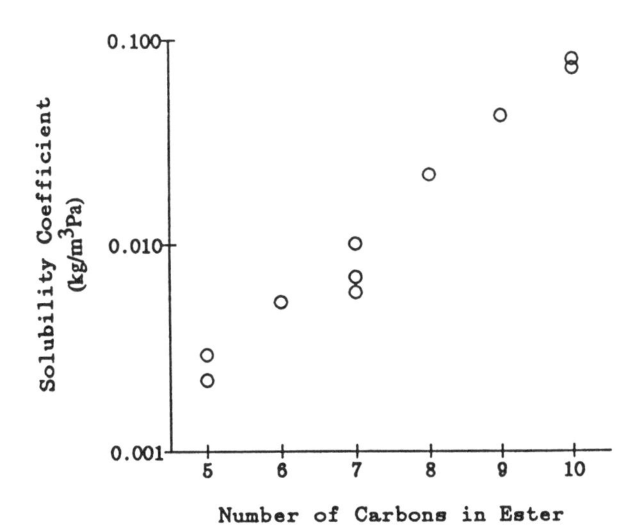

Figure 4. Solubility coefficients of linear esters at 85 °C in a vinylidene chloride copolymer film. (Reproduced with permission from ref. 7. Copyright 1990 American Chemical Society.)

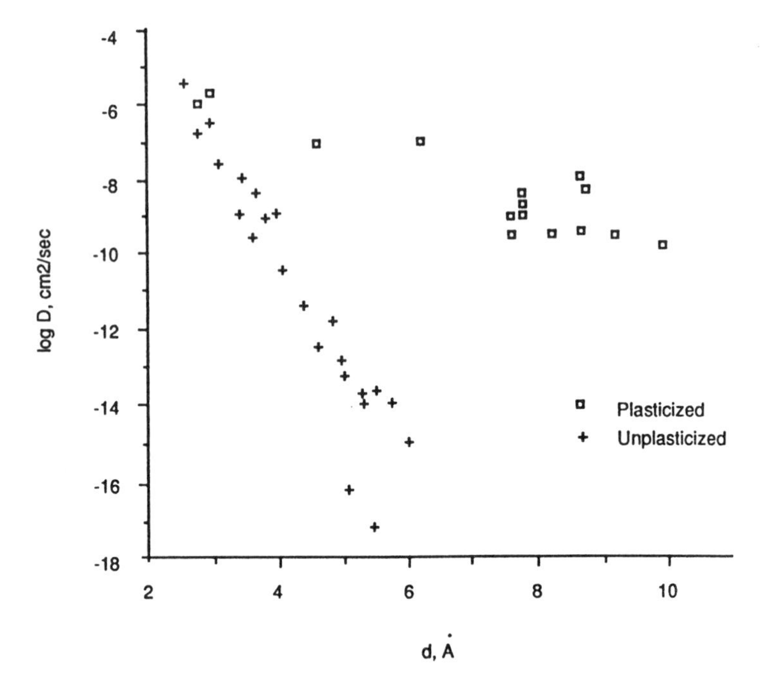

Figure 5. Diffusion coefficients of small molecules in poly(vinyl chloride) at 30 °C. (Adapted from ref. 9. Copyright 1990 American Chemical Society.)

diffusant size (*13-15*). Diffusion coefficients for molecules with the sizes of flavor molecules can not be measured accurately because they are too low.

The author has worked with thin Nylon 6 films with limited success. Only by working at elevated temperatures and with smaller molecules could diffusion coefficients be measured. Using room temperature or real flavor molecules did not yield quantitative data.

Summary

The component parts of the permeability, namely the diffusion coefficient and the solubility coefficient, are important to understanding the interaction of flavors with plastic packaging.

The diffusion coefficients in polymers are functions of the size of the flavor. Larger flavors have smaller diffusion coefficients. The dependency is so strong for glassy polymers that diffusion coefficients for flavors are so small that they can not be measured accurately.

The solubility coefficients in polymers are very large. This means that a plastic package can absorb a great deal of flavor if the diffusion coefficient is large enough for the penetration to occur during the storage time of the product.

Literature Cited

1. Landois-Garza, J.; Hotchkiss, J. H. In *Food and Packaging Interactions*; ACS Symposium Series 365; Hotchkiss, J. H., Ed.; ACS Symposium Series 365; American Chemical Society: Washington, D.C., 1987, pp. 42-58.
2. DeLassus, P. T. *J. Vinyl Technology* **1979**, *1*, 14-19.
3. DeLassus, P. T. *J. Vinyl Technology* **1981**, *3*, pp. 240-245.
4. Salame, M. *J. Polym. Sci.* **1973**, Symposium No. 41, 1-15.
5. Van Krevelen, D. W.; *Properties of Polymers*, 3rd ed., Elsevier: Amsterdam, The Netherlands, 1990, Chapter 18, pp. 535-583.
6. DeLassus, P. T.; Tou, J. C.; Babinec, M. A.; Rulf, D. C.; Karp, B. K.; Howell, B. A. In *Food and Packaging Interactions*, Hotchkiss, J. H., Ed.; ACS Symposium Series 365; American Chemical Society: Washington, D.C., 1987, pp. 11-27.
7. Strandburg, G.; DeLassus, P. T.; Howell, B. A. In *Barrier Polymers and Structures*, Koros, W. J., Ed.; ACS Symposium Series 423; American Chemical Society: Washington, D.C., 1990, pp. 333-350.
8. DeLassus, P. T.; Strandburg, G.; Howell, B. A. *Tappi Journal* **1988**, *71*, 177-181.
9. Berens, A. R. In *Barrier Polymers and Structures*, Koros, W. J., Ed.; ACS Symposium Series 423; American Chemical Society: Washington, D.C., 1990, pp. 92-110.
10. Zobel, M. G. R. *Polymer Testing* **1985**, *5*, 153-165.
11. Stannett, V. In *Diffusion in Polymers*, Crank, J.; Park, G. S., Eds.; Academic Press: London, 1968, pp. 41-73.
12. Fujita, H. In *Diffusion in Polymers*, Crank, J.; Park, G. S., Eds.; Academic Press: London, 1968, pp. 75-105.
13. Berens, A. R.; Hopfenberg, H. B. *Journal of Membrane Science* **1982**, *10*, 283-303.
14. Chen, S. P., *Polymer Preprints* **1974**, *15*, 77-81.
15. Edin, J. A. D.; Chen S. P. *Organic Coatings and Plastics Chemistry* **1978**, *39*, 248-253.

Measurement Tools

Chapter 15

Gas Chromatography–Olfactometry as a Tool for Measuring Flavor–Food Ingredient Interactions in Model Systems

Norbert Fischer[1] and Tony van Eijk[2]

[1]Research Division and [2]Flavor Application Department, Dragoco AG, D–37601 Holzminden, Germany

Changes in flavor profile that depend on food matrix variations are often studied by measuring the flavor release in model systems by means of headspace gas chromatography (HS-GC) techniques. The analyst relies on the quantification of volatiles in the headspace. For many flavor components of high sensory potency, however, gas chromatography-olfactometry (GC-O) represents the only useful detection method, since the concentrations usually encountered in the headspace above foods are too low to be quantified or even detected by the instrument as flame ionization detector (FID) peaks. The incorporation of the GC-O or "GC-sniffing" technique into the headspace analysis helps to identify and quantify important trace constituents in complex flavors, and improves the correlation with the sensory profiles of the complete model system. We report here experiments to characterize flavor changes in model emulsions, based on headspace-GC and GC-O methods.

The phrase "flavor–food ingredient interactions" comprises many aspects of the different effects that "bulk" food constituents can have on flavor perception. If phenomena such as irreversible flavor binding are set aside, then the major influence of matrix constituents on flavor is control of the distribution of flavor compounds between the "food" and "gas phase", and hence their release behavior.

The measurement of flavor release, with its intensity-related and temporal components (*1,2*) can be approached using a combination of sensory and analytical methods. On the one hand, sensory methods (descriptive sensory analysis, time-intensity measurements) are applied to describe and quantify specific flavor attributes as they are influenced by the complex food. Flavor release behavior, on the other hand, can be investigated by analyzing the headspace composition above a given food sample.

0097–6156/96/0633–0164$15.00/0

The simulation of retronasal flavor perception (temperature of the mouth cavity, incorporation of mixing and shear forces, etc.) in suitably designed model vessels allows one to understand the behavior of a complex flavor in a food as it is eaten. Such simulation is useful because the concentrations of individual volatiles in the headspace above a food sample as determined, e.g., by mass spectrometry (MS) (*3*), differs from the distribution of volatiles in the "mouthspace", which can be measured, e.g., by the "MS-breath method" (*4*).

Since the complexity of real food systems makes it very difficult to obtain meaningful experimental data and interpretations, it is useful to approach investigations of flavor–food ingredient interactions by reducing "real life" complexity to simplified and more controllable model systems (see e.g., *5-7*). These model systems allow the study of the effects of individual matrix constituents on the flavor release behavior of individual flavor components.

From a commercial point of view, an important variable in a food matrix today is the fat content, which quite often is reduced to make foods "healthier", relative to the conventional full-fat versions.

Fat, as a good "solvent" for flavor components has a major influence on the partitioning of flavors between the "food" and "gas phase" and hence on flavor perception (*8-10*). Experience shows that flavorings designed for aqueous systems perform poorly in fat-containing systems, and flavorings designed for fat-based systems tend to become unbalanced or even off-flavored in aqueous or reduced-fat systems (*11-12*).

The interest of a flavor house in evaluating flavor - food interactions arises from the necessity to develop (and sell) flavors that are optimized for different food systems. Any change in a food matrix dictates a modification of the flavor in order to optimize its performance. Quite often flavorings have to be tailored to a customer-supplied base, and sometimes matrix-based off-flavors which were hidden in the original food version by the high fat-content have to be masked by the flavor. Flavor changes that occur with food matrix variations have to be evaluated on a sensory basis and in relation to individual flavor components; this means that analytical data on the headspace composition and sensory profiles obtained by descriptive analysis have to be correlated in order to be able to reformulate flavors for different matrices.

In our work, we started to investigate the headspace flavor composition above model food systems at physiological temperature (37 °C), thus simulating in a very simplified way the flavor release in the mouth.

Headspace Methodology

The measurement of individual headspace volatiles can be performed under static or dynamic conditions. The former measures the concentration of volatiles under equilibrium conditions (*13*), while the latter to the kinetics of flavor release (*3*) and therefore to the temporal aspects of flavor perception. Consequently, the appropriate headspace technique must be selected based on the application (see Table I). If the equilibrium concentrations need to be measured, static headspace would be chosen. However, static headspace techniques suffer from a major drawback: "The static

Table I. Comparison of Static and Dynamic Headspace Methods

Static Headspace	*Dynamic Headspace*
Equilibrium conditions	Non-equilibrium conditions
Low sensitivity for detection	Higher sensitivity
Enrichment step possible	Enrichment necessary

headspace technique fails when trace components or components with very low vapour pressure are analyzed" (*14*). Where headspace analysis of flavors is concerned, this drawback can be compensated by using the nose as a more sensitive bioasay to detect relevant trace constituents. In addition to increasing sensitivity, GC-O is an indispensable prerequisite for discriminating between relevant (flavor active) and non-flavor active volatiles. "Flavor" is not simply the sum of "volatiles" that can be measured e.g., by means of GC-FID, but rather a subset of the sensorially-relevant volatiles (*15*).

GC-O, therefore, provides an important additional detection tool in flavor research, and experience shows that many key trace components (*16*) that cannot be quantified by GC-FID or GC-MS can be detected by the GC-O bioassay (See Table II).

Table II. Threshold Values of Some Selected Flavor Components of High Sensory Importance

Component	*Odor Description*	*Threshold Value*	*Literature Ref.*
4-Methoxy-2-methyl-2-butanethiol	Blackcurrant	0.03-0.06 ppb (oil)	(*17*)
β-Damascenone	Warm-fruity	0.002-0.009 ppb (H_2O)	(*18*)
1-Octen-3-one	Mushroom	0.005-0.1 ppb (H_2O)	(*18*)
Ethyl- 2-methyl-butanoate	Apple-like	0.1-0.3 ppb (H_2O)	(*18*)

In this study, the application of static headspace-GC and headspace-GC-O (HS-GC-O) to the evaluation of changes in the flavor profile is demonstrated. Emulsions of varying fat content were flavored with a model "red berry" flavoring to serve as a model food system.

The kinetics of flavor release, which influence the temporal aspects of flavor perception, are not considered within this study.

Experimental

The experimental set-up was kept as simple as possible: as a vessel to simulate flavor release in the mouth, an Erlenmeyer flask (50 mL) equipped with a septum head was used. The release vessel together with the syringes used in the experiments were placed in a thermostated environment (Uniequip incubator hood, 50 x 50 x 60 cm, temperature 37 °C); transfer of the headspace volatiles into the GC was accomplished by means of a (pre-warmed) gas-tight syringe (10 mL).

Composition / Preparation of Model Emulsions. Following are the formulas used for preparation of the model flavor and emulsion systems:

- Model "red berry" flavoring in triacetin 0.1 to 0.5% in emulsion
- Emulsifier: mono-/ diglyceride citrate 1% (w/v)
- Emulsions: (Both procedures showed very little difference with regard to the headspace composition.)
    - a) Flavor cocktail dissolved in water / oil + emulsifier added (60 °C), emulsification by means of an Ultraturrax high-speed mixer (4000 min^{-1}, 30 sec.)
    - b) "Neat" emulsion prepared as in a), flavor cocktail added after cooling to room temperature, equilibration overnight.

Chromatographic Conditions. The following conditions were used: GC Siemens Sichromat 1-4, parallel detection FID/sniff port; heated flexible transfer line to sniff port (180 °C); sniffing mask purged by a stream of humidified air (approx. 100 mL/min); column: 30 m x 0.32 mm DB-1 (1 mm film); cryofocussing (*7,19*): first column loop immersed in liquid N_2; after injection of headspace sample (injection speed ca. 10 mL in 30 sec.), liq. N_2 removed, oven closed, start temperature program (40 °C at 4°/min to 240 °C). To the individual odor notes perceived during GC-O, a subjective intensity was assigned: 1 = weak, 4 = strong; experiments were repeated at least once, and most were repeated by different sniffers.

Results and Discussion

The model "red berry" flavoring was dissolved in water and pure oil, incorporated into model emulsions with a fat content ranging from 1% to 80%, and in real food systems such as whole milk (3.5% fat) and cream (24% fat). The concentration of the total volatiles in the headspace (FID-detection) is compared to the total sensory intensities, as perceived during HS-GC-O in Figure 1. At first inspection, a correlation between the total volatile concentration and the total sensory intensity seems apparent; both variables decrease with increasing fat content, as expected. However, Table III shows that the total sensory intensity is distributed between five individual flavor components, of which only ethyl 2-methylbutanoate is detectable by means of GC-FID (see chromatogram in Figure 2).

Table III also demonstrates that the components of this flavor are influenced differently by the increase in fat content in the model systems: gamma-decalactone and beta-ionone are perceived in the pure water sample, but are suppressed below their

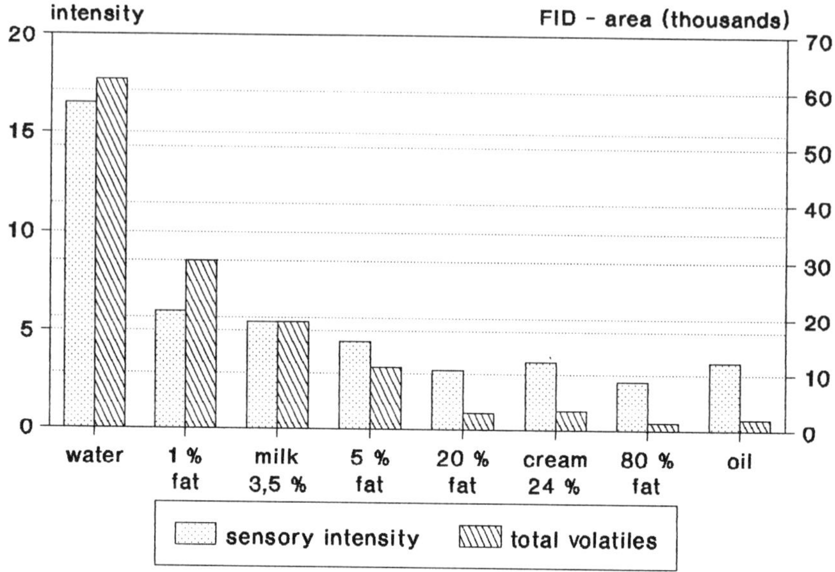

Figure 1. Comparison of total sensory intensities (GC-O) and total volatiles (FID detection) for model emulsions and milk products.

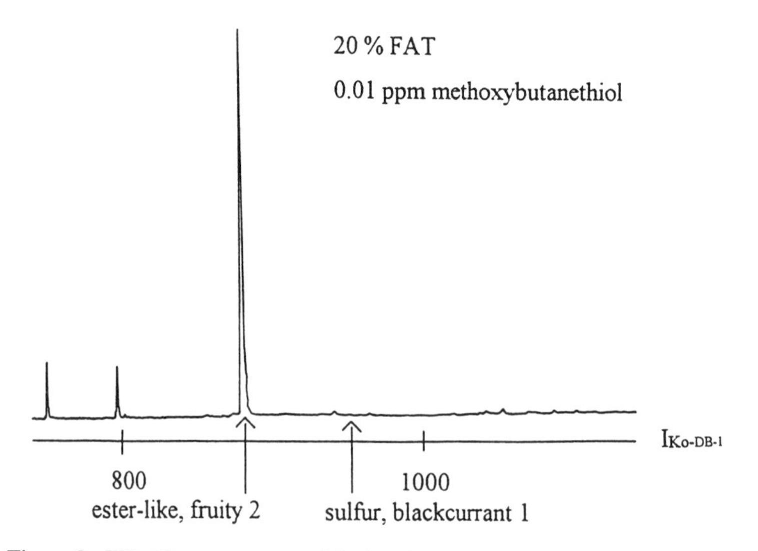

Figure 2. FID-Chromatogram of the headspace from a model emulsion (20% fat); comparison of the sensory impressions from GC-O of a trace component (4-methoxy-2-methyl-2-butanethiol) and a major component (ethyl 2-methylbutanoate).

Table III. Intensity of Important Components of "Red Berry" Model Flavoring in Different Model Food Systems (HS-GC-O)

Component	*Sensory Intensity (GC-O)[1] in*					
	water	*1% fat emuls.*	*milk 3.5%*	*5% fat emuls.*	*20% fat emuls.*	*oil*
Ethyl 2-methylbutanoate	4	4	4	4	2-3	3
4-Methoxy-2-methyl-2-butanethiol	3	1	1	<1	<1	<1
β-Damascenone	3	1	<1	-	-	-
γ-Decalactone	2-3	-	-	-	-	-
β-Ionone	4	-	-	-	-	-

[1] Subjective intensity scale: 1 = weak, 4 = strong; - = not recognizable.

detection threshold even in the 1% fat emulsion. The different behaviour of these impact components also implies that the profile of the model flavor changes dramatically with variation in fat content; the sensory descriptions for three of the model emulsions given in Table IV corroborate this.

Table IV. Sensory (Odor) Descriptions of Flavored Model Emulsions (0.1% Flavoring)

Emulsion	*Odor Description*
1 % Fat	Fresh-fruity, blackcurrant-like, herbaceous
5 % Fat	Fruity-sweet, raspberry-like, slightly green-leafy
20 % Fat	Fruity, sweet, strawberry, peach-like, creamy, malty[1]

[1] The model flavoring also contained maltol and 2,5-dimethyl-4-hydroxy-3(2H)-furanone, which are responsible for the malty notes occurring in the 20% fat sample, but could not be measured by HS-GC-O due to their low volatility / high water solubility.

As an example of a high-impact flavor component that can only be detected by HS-GC-O, 4-methoxy-2-methyl-2-butanethiol, a well-known character impact component of blackcurrant (*20*), was incorporated into the model flavor. In Figure 3, the sensory intensities and FID peak areas of "methoxybutanethiol" and ethyl

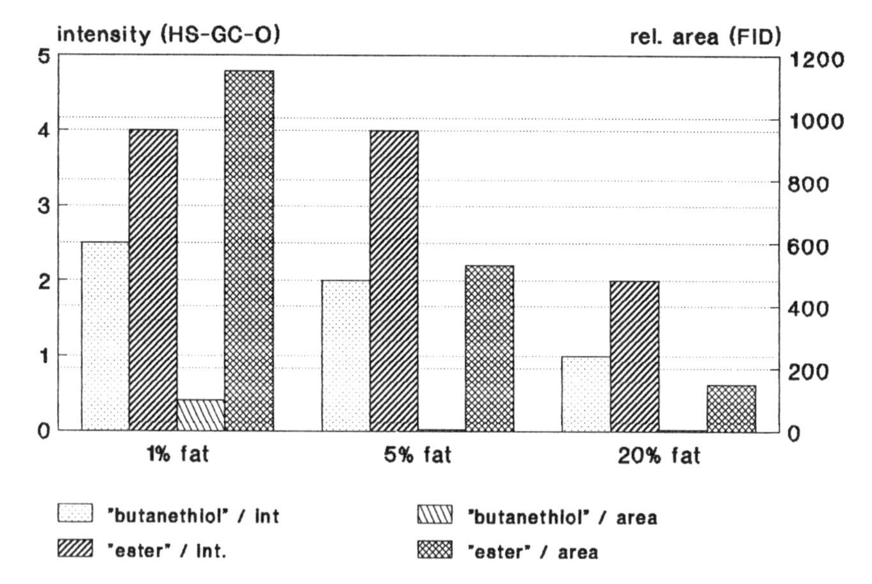

Figure 3. Sensory intensity versus peak area for ethyl 2-methylbutanoate and 4-methoxy-2-methyl-2-butanethiol.

2-methylbutanoate are compared in different model emulsions (O/W). It can be seen that in the 5 and 20% fat samples no FID peak was obtained for the methoxybutanethiol, but nevertheless an intensity ratio 2 and 1, respectively, was observed.

If the flavor intensity of the methoxybutanethiol had to be adjusted in order to adapt the flavor profile from a 1% fat matrix to a 20% fat matrix, then a "correction factor" could be derived by simply comparing the perceived sensory intensity of the compound during HS-GC-O. This procedure, however, would only be based on a subjective intensity rating. This leads to reliable relative rankings when two samples are compared by only one test person.

To more accurately quantify these differences in sensory intensity, the technique of aroma extract dilution analysis (*21*) or *CHARM* analysis (*22*) can be adapted to the headspace measurements. An analogue to the "dilution factor" can be obtained by simply injecting decreasing headspace volumes from the flavor release vessel, until no more odor impression at a given retention time (retention index range) is perceived. A similar procedure has been described for the identification of potent odorants in the headspace above tea powders (*23*). Table V shows the application of the headspace "dilution" analysis to the quantification of the sensory intensity of the methoxybutanethiol in 1% and 20% fat emulsions; in the low-fat system, injection of 1 mL headspace is sufficient to recognize the "cassis-note" during GC-O, while 3 mL of the headspace of the 20% fat-system had to be injected to achieve sensory detection. A comparison of the two "dilution" factors obtained – 10 to 3.3 – would lead to a "correction factor" of about three for adaptation of the methoxybutanethiol intensity from the 1% fat system to the 20% fat system.

If we go back to the subjective sniffing intensities assigned to the methoxybutanethiol peak in these two samples (Table III), then an intensity ratio of 2.5 (2.5 divided by 1) would be obtained, compared to a correction factor of 3 obtained from the "dilution" analysis, which demonstrates that the direct comparison of a given odor attribute by a trained "sniffer", using a subjective intensity rating, also leads to useful results.

Another example that demonstrates the usefulness of the HS-GC-O technique for evaluating flavor changes involves detection of a high-impact off-flavor component (in this example, 1-octen-3-one) that becomes sensorially relevant in a flavor profile when the fat content of the matrix is lowered. This example represents the frequent situation that performance of a flavoring is hampered by an off-flavor component which emerges in a low-fat food system, but was masked by the fat in the original full-fat product.

In Figures 4 and 5, the peak areas and sensory intensities of 1-octen-3-one peak are compared for pure water or pure oil systems and for emulsions with 1.5 and 20% fat. A reasonable FID-peak for the octenone could only be obtained for the pure water sample. Sensory detection, however, revealed the presence of an off-flavor (the mushroom-like and metallic note typical for 1-octen-3-one) for all samples except the 20% fat system.

If, according to this example, a 20% fat food matrix would be changed in order to give a low-fat product with 1% fat, then a mushroom-like off-flavor would be

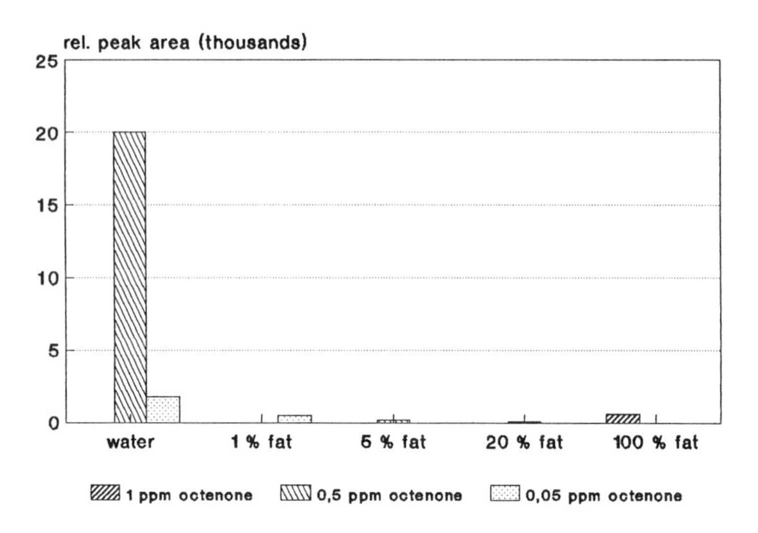

Figure 4. Peak area (FID detection) of 1-octen-3-one in the headspace above model systems of varying fat content. Headspace injected: 10 mL. 1-Octen-3-one (1 ppm level) measured only at 100% fat level; 0.5 ppm measured in water, 5% and 20% fat emulsion; 0.05 ppm level measured in water and in 1% fat emulsion.

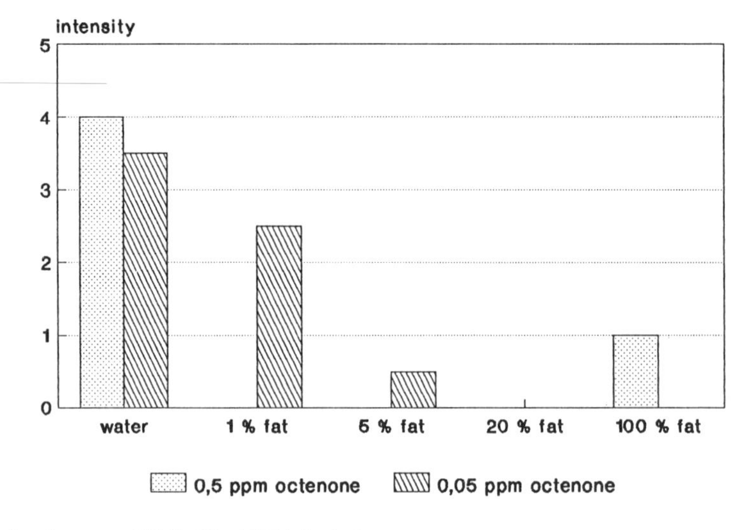

Figure 5. Sensory intensity (GC-O) of 1-octen-3-one in the headspace above model systems of varying fat content. Injection volume and dosages of 1-octen-3-one as in Figure 4.

observed, which could be measured by applying the HS-GC-O methodology to this food system.

Table V. Static Headspace "Dilution" Analysis of 4-Methoxy-2-methyl-2-butanethiol in 1% and 20% Fat Emulsions

	Headspace[1] (mL)	*"Dilution" Factor[2]*
1 % Fat emulsion	1	10
20 % Fat emulsion	3.3	3

[1] Volume of headspace (mL) needed for obtaining a sensory impression during GC-O.
[2] Calculated relative to the "normal" injection volume of 10 mL headspace for GC-O detection.

Possible Correlation with Descriptive Sensory Analysis

It is important to note that HS-GC-O, as a sensory method that relates to *individual flavor components*, is *complementary* to descriptive sensory analysis as a technique that evaluates *overall flavor attributes*. Therefore, the HS-GC-O of a flavoring in the food matrix can be used to determine descriptors for sensory analysis, specifically for those descriptors that are correlated with components of high intensity in GC-O. The evaluation of the applicability of combined HS-GC-O and descriptive sensory analysis, is part of our ongoing studies in this field.

Detection Sensitivity of the HS-GC-O Method

In the model experiments described, a higher dosage (10-20X) had to be applied for the HS-GC-O measurements than for sensory evaluation in order to facilitate detection of a reasonable number of components in the flavor cocktail. There are two main reasons for this reduced "sensitivity" for individual components in HS-GC-O:

1) A 1:1 split between FID and sniffing port was used for the gas chromatography which accounts for 50% of the "sensitivity loss".
2) The observed total flavor intensity of the sample is separated during GC into a set of individual flavor components, each of which is less intense than the total mixture. This argument is based on the assumption that, as a first approximation, individual flavor components exhibit an additive behaviour in a mixture with respect to their flavor intensity.

There are several ways to increase the sensitivity for individual flavor components at the sniffport:

Figure 6. Correlation between percentage of headspace injected and peak area (FID detection) calculated for ethyl 2-methylbutanoate. Headspace vial volume is 70 mL.

- an elevated concentration of the total flavoring can be applied (this was done for the experiments described here),
- the simultaneous FID-detection could be abandoned, and
- the transfer of headspace components into the gas chromatograph can be optimized.

Optimization of the transfer of components involves increasing the amount of volatiles injected into the GC-O system. To simply increase the percentage of headspace that is drawn from the release vessel, however, is only of limited use. As soon as the headspace volume removed is no longer negligibly small relative to the volume of the sample vessel, a non-linear relationship is obtained between "volume injected" and "total peak area" (see Figure 6). Since outside air is entering the system during the removal of headspace using a syringe, the headspace is continuously diluted, and the "complete" transfer of the volatiles from the headspace into the GC would require an indefinite sample volume. At the same time, "static" headspace conditions cannot be maintained and "dynamic" headspace conditions ensue. In our system, the non-linearity became obvious at injected headspace volumes above ca. 10% of the vessel volume.

We therefore tried to develop a "headspace displacement" procedure which would allow for the "whole" headspace above a food sample to be transferred into the gas chromatograph while maintaining static headspace conditions. A similar approach was useful for the determination of odor thresholds using squeezable teflon bottles (*9*); this technique delivers the headspace vapor above a flavoring solution to the nose with minimum dilution from outside air.

In its initial form, the system is comprised of two gas syringes (see scheme in Figure 7) coupled by a valve, both placed in the thermostated incubation hood; the first syringe represents the "flavor release vessel" and contains the model emulsion, a stirring bar and a defined headspace volume above the solution. After reaching equilibrium, the headspace can be transferred into the second syringe by simultaneously moving both syringe plungers. This avoids changing the headspace volume, and thereby maintains equilibrium conditions. After closing the valve, the syringe can be removed, fitted with an injection needle and the headspace sample can be injected into the GC.

Initial results demonstrated the suitability of this optimized "static headspace displacement" technique; a linear relationship between "percentage of headspace injected" and "FID peak area" can be obtained for percentages of up to more than 90% of the headspace volume (compare with Figure 6).

The transfer of larger headspace volumes into the GC requires the application of more advanced intermediate trapping techniques (see e.g. *24*) rather than a simple cryofocusing of the headspace volatiles in a cooled column loop, e.g., injection into a liquid nitrogen-cooled programmable temperature vaporizer (PTV), in which the volatiles are trapped on a cooled adsorbent. The main advantage of this enrichment technique is injection of the headspace sample with inlet split opened, which avoids the issue of overcoming the column backpressure during injection of the sample volume.

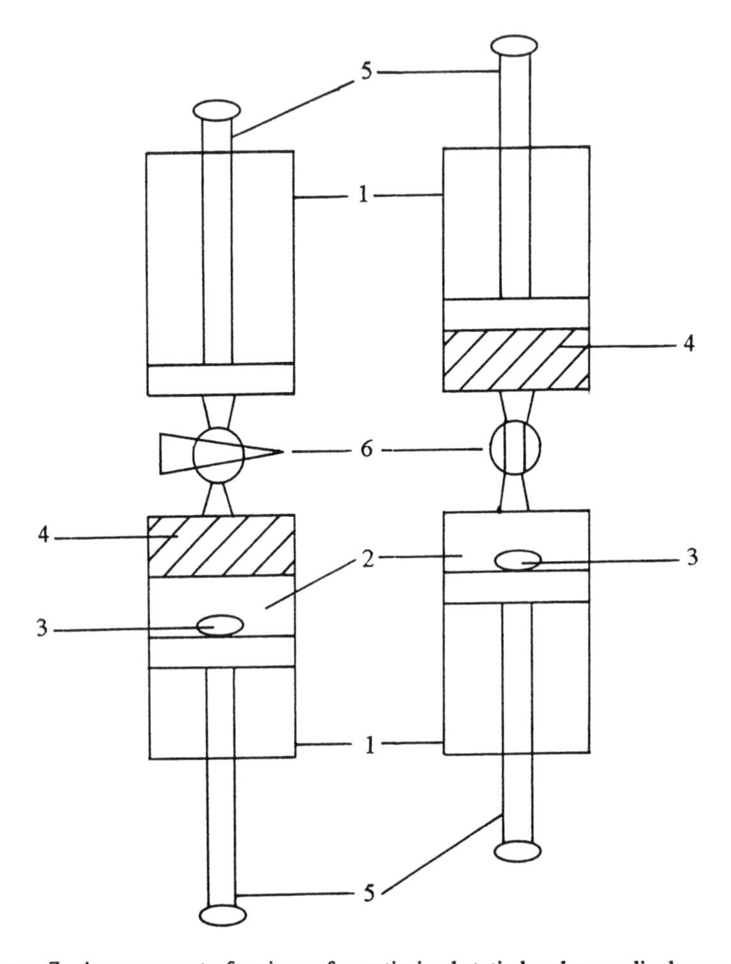

Figure 7. Arrangement of syringes for optimized static headspace displacement: (1) gastight syringe, 50 mL, (2) liquid sample, (3) stirring bar, (4) headspace gas, (5) plunger, (6) valve.

An optimized static headspace transfer system for HS-GC-O connected with the flavor release vessel should represent a useful instrumental set-up for the measurement of the release of flavor components from food systems.

Conclusion

The aim of our investigation was to establish a technique to measure the release behaviour of a flavor in every desired complex food matrix by means of headspace-GC-olfactometry, and thus to be able to optimize the performance of a flavoring in any given food by using this technique in conjunction with descriptive sensory analysis. We have successfully demonstrated the application of headspace-GC-O to quantify trace constituents in low- and high-fat model emulsion systems.

Acknowledgment

The authors would like to thank Ellen Gruber and Annegret Mönnikes for their valuable technical support.

Literature Cited

1. Overbosch, P.; Afterof, W. G. M.; Haring, P. *Food Rev. Int.* **1991**, *7*, 137-184.
2. Lee, W. E. III; Pangborn, R. M. *Food Technol.* **1986**, *40*, 71-78, 82.
3. Lee, W. E. III. *J. Food Sci.* **1986**, *51*, 249-250.
4. Soeting, W. J.; Heidema, J. *Chem. Senses* **1988**, *13*, 607-617.
5. Schirle-Keller, J.-P.; Chang, H. H.; Reineccius, G. A. *J. Food Sci.* **1992**, *57*, 1448-145.
6. Bakker, J.; Salvador, D.; Langley, K. R.; Potjewijd, R.; Martin, A.; Elmore, J.S. *Poster presented at Int. Conference "Understanding Flavour Quality"*, Bristol, UK, Sept. 20-23, 1992.
7. Ebeler, S. E.; Pangborn, R. M.; Jennings, W. *J. Agric. Food Chem.* **1988**, *36*, 791-796.
8. Forss, D. A. *J. Agric. Food Chem.* **1969**, *17*, 681-685.
9. Buttery, R. G.; Guadagni, D. A.; Ling, L. *J. Agric. Food Chem.* **1973**, *21*, 198-201.
10. LeThanh, M.; Pham Thi, S. T.; Voilley, A.. *Sci. Aliments* **1992**, *12*, 587-592.
11. Hatchwell, L. C. *Food Technol.* **1994** *(2)*, 98-102.
12. Plug, H.; Haring, P. *Trends Food Sci. Technol.* **1993**, *4*, 150-152.
13. Guth, H.; Grosch, W. *Flav. Fragr. J.* **1993**, *8*, 173-178.
14. Nunez, A. J.; Gonzalez, L. F.; Janak, J. *J. Chromatogr.* **1984**, *300*, 127-162.
15. Acree, T. E. In *Flavor Science, Sensible Principles and Techniques*, T. E. Acree, R. Teranishi, Eds., American Chemical Society: Washington, DC, 1993, p. 1-20.
16. Grosch, W. *Trends Food Sci. Technol.* **1993**, *4*, 68-73.
17. Boelens, M. H.; van Gemert, L. J. *Perfum. Flavor.* **1993**, *18 (3)*, 29-39.
18. Leffingwell, J. C.; Leffingwell, D. *Perfum. Flavor.* **1991**, *16 (1)*, 1-19.
19. Macku, C.; Kallio, H.; Takeoka, G.; Flath, R. *J. Chromat. Sci.* **1988**, *26*, 557-560.

20. Rigaud, P.; Etievant, P.; Henry, R.; Latrasse, A. *Sci. Aliments* **1986**, *6*, 213-220.
21. Ullrich, F.; Grosch, W. *Z. Lebensm. Unters. Forsch.* **1987**, *184*, 277-282.
22. Acree, T. A.; Barnard, J.; Cunningham, D. *Food Chem.* **1984**, *14*, 273-286.
23. Guth, H.; Grosch, W. *Flavour Fragr. J.* **1993**, *8*, 173-178.
24. Nitz, S.; Jülich, E. In *Analysis of Volatiles*, P. Schreier, Ed., Walter de Gruyter, Berlin, 1984, pp. 151-170.

Chapter 16

Retronasal Flavor Release in Oil and Water Model Systems with an Evaluation of Volatility Predictors

Deborah D. Roberts[1] and Terry E. Acree

Department of Food Science and Technology, Cornell University, New York State Agricultural Experiment Station, Geneva, NY 14456

A device was constructed to simulate retronasal aroma incorporating the conditions of salivation, stirring, and air flow that occur in the mouth. Made from a 4-L blender, the simulator was able to handle large sample sizes for increased sensitivity. The volatilities of eight flavor compounds in soybean oil or water were compared, with a calculation of the first-order rate constant, k (min^{-1}). The most volatile compounds in water, α–pinene, ethyl-3-methyl butyrate, and 1,8-cineole had large rate constants of 3.3×10^{-1}, 1.4×10^{-2}, and 2.3×10^{-3} min^{-1}, respectively, which decreased by two to four orders of magnitude in oil. Two moderately volatile compounds, 2-methoxy-3-methyl pyrazine and butyric acid, behaved differently from each other in oil. Volatility of the pyrazine decreased 7-fold in oil while the acid showed no decrease. The highly non-volatile compounds, methyl anthranilate, maltol, and vanillin, showed low volatilies in both systems. Log P (octanol-water partition coefficients) were measured for the compounds and related to their oil versus water volatilities. Several chemical parameters of the compounds were evaluated for their prediction of volatilities and found to only roughly correlate.

As food is eaten, it is the retronasal passage of aroma volatiles, through the nasopharanx, which causes the odor component of the flavor sensation. Orthonasal aroma (inhalation through the external naris) and retronasal aroma have been shown in to be different in quality and quantity. This is probably because of different conditions of the food in the mouth, as well as different paths of odorants to the nasal cavity. Some studies compared only the pathway and not the effects of the mouth conditions by inhaling the odor of a food through a straw. The intensity of inhaled retronasal aroma was found to be less than orthonasal aroma for meat

[1]Current address: Nestlé Research Center, Lausanne, Nestec Ltd., Vers-Chez-Les-Blanc, 1000 Lausanne 26, Switzerland

0097–6156/96/0633–0179$15.00/0

flavoring, citral and vanillin (*1, 2*). However, when the samples were actually placed in the mouth, retronasal aroma had a lower threshold for citral and vanillin than orthonasal aroma (*2*) and both had different slopes of Stevens' regression lines. In addition, the temporal pattern of sensation for citral and vanillin was found to be different as retronasal aroma had a less intense and longer temporal pattern than orthonasal aroma (*3*). Salivation and mastication in the mouth are probably responsible for differences in flavor volatility for ortho- and retronasal aroma. A benchmark device has thus been constructed which measures the retronasal volatility of food by incorporating the elements of mastication and stirring. This study describes the device and preliminary experiments to study the volatility of flavors in oil and water model systems. The models were combined with simulated saliva. In the preliminary experiments reported here, an appropriate amount of water was used to simulate saliva. More complex simulations are reported elsewhere (*4*). Known and measured parameters of the octanol-water partition coefficient and vapor pressure of the pure flavor compound were evaluated as predictors of the retronasal flavor volatility.

Materials and Methods

Samples. The particular flavors were chosen because they are important contributors to a variety of foods and they exhibit a range of chemical properties. An ethanolic flavor mix (2 mL) of ethyl-2-methyl-butyrate, α–pinene, and 1,8-cineole, maltol, vanillin, butyric acid, methyl anthranilate and 2-methoxy-3-methyl pyrazine (Aldrich, Milwaukee, WI) was added to 2 L of soybean oil and water. The flavor levels corresponded to the amount found naturally in a food product characteristic of that aroma in order to ensure that the apparatus would have the sensitivity to analyze the flavor in an actual food. The amounts were 0.1, 6.0, 15, 19, 2.0, 15, 4.0, and 11.0 mg/L, respectively.

Retronasal Aroma Simulator. A 4-liter Waring blender (Figure 1) formed the basis for the apparatus. An air inlet entered the base of the blender with 1/8" copper tubing and Swagelok joints which was monitored by a Brooks-Mite Flow Indicator. The air outlet at the top was connected to 1/8" teflon tubing by Swagelok joints, which joined a Waters silica Sep-Pak via a Bio-Rad 3-way luer-lock stopcock. Silica Sep-Paks were activated by heating at 125 ^{o}C for 16 hours just prior to the experiment. A Superior Electric Volt Box auto-transformer was used to control the speed and on/off switch of the blender. The blender speed was on low, with an auto transformer set to 60 volts which gave 300 rpm. The air flow into the blender was 1890 mL/sec.

After the flavor mix was added to the sample, the 2-L oil and water matrices were stirred for 30 and 3 minutes, respectively. After the addition of 500 mL water (simulated saliva) to the retronasal aroma simulator, the blender lid was tightly shut, the air and mixing were turned on, and collection of volatiles on a silica Sep-Pak began. After 2.5 minutes, a new Sep-Pak was attached. In total, six Sep-Paks were used. The six Sep-Paks were used to observe the dynamics of the flavor release. After elution of the Sep-Paks with 4 mL of redistilled ethyl acetate, the levels of the flavor compounds were quantitated by GC/MS and GC/FID using an internal standard curve.

Measurement of Partition Coefficients. The partition coefficient of a compound between octanol and water (P):

$$P = \frac{\text{Concentration in Octanol Phase}}{\text{Concentration in Water}}$$

The measurement of log P for all of the flavor compounds was performed by the traditional shake-flask method at 25 ^{o}C. An initial octanol starting concentration of about 10 mg/mL was used for all of the flavors except maltol. Because of maltol's low solubility in octanol, a level of 2 mg/mL was used. A 3-mL aliquot of the flavor in octanol was shaken for 1 minute on a rotary evaporator with 3 mL of water. After centrifugation, a sample from the octanol phase was diluted in ethyl actate and then quantified using an internal standard curve on the GC/MS and GC/FID. Three replications of the shake-flask method were made. The beginning octanol concentration was analyzed in triplicate and each of the three final octanol concentrations was analyzed in duplicate. The level of flavor in water was calculated by the difference between the initial and final concentration in octanol. However, for pinene and methyl anthranilate, the amounts in octanol before and after shaking were not significantly different so the water was analyzed using ethyl acetate extraction.

Results and Discussion

Design Justification for the Retronasal Aroma Simulator. The large volume of the Retronasal Aroma Simulator (Figure 1) allowed the collection of sufficient volatiles for analysis of trace components. After desorption from the volatile trap, the solvent could be concentrated, if necessary, to improve sensitivity. In order to simulate mouth conditions sufficient to establish a benchmark, the design included saliva, chewing and crushing conditions, and air flow dynamics similar to that occuring during eating.

During mastication, mixing of food occurs as force is applied by the tongue and teeth, causing shear stress, and the break-up of food. Shear rate or the velocity gradient established in a fluid as a result of an applied shear stress (*5*) has been measured in the mouth during eating. The shear rate was not constant but varied from 10-500 s^{-1} depending on the food, the individual, and the point in the process of mastication (*6-8*). The shear rate in Retronasal Aroma Simulator was estimated from rotational speed and impeller size (*9*) and found to be about 30 s^{-1}.

The reported sniff volume flow rate during inhalation is about 100 mL/sec (*2*). In retronasal aroma, however, the gas that adsorbs the volatiles from food is not directly in line with the respiratory air flow. The volume flow rate over food in the mouth must be much lower than 100 mL/sec. The apparatus, which accepted a

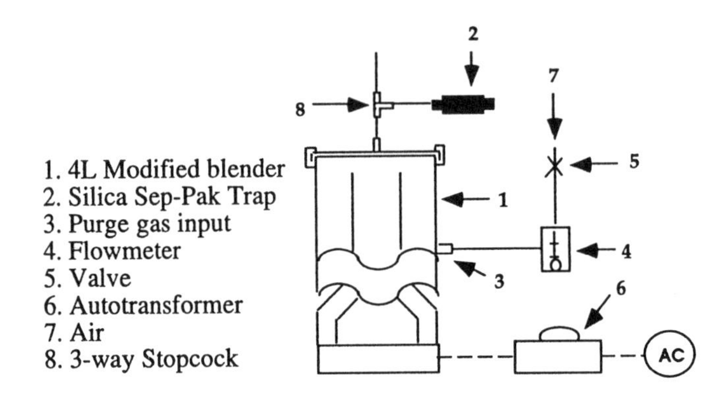

Figure 1. Schematic diagram of the retronasal aroma simulator.

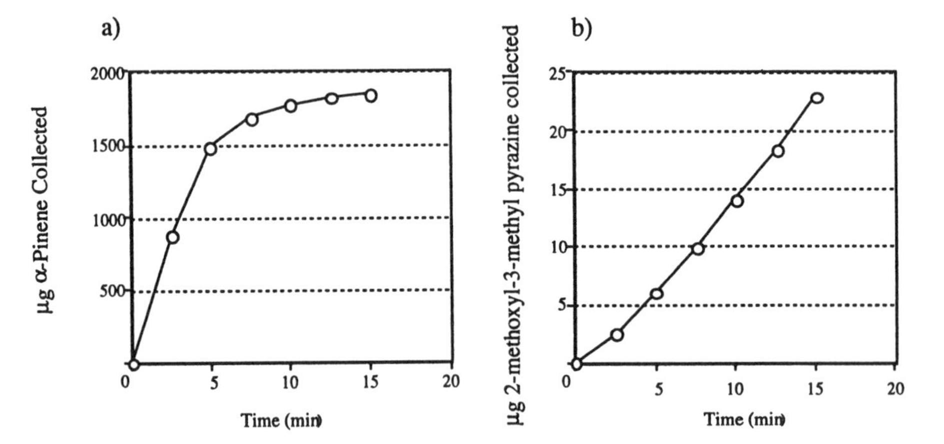

Figure 2. Volatilization curves for flavor compounds in water, showing the amount collected by Sep-Paks over time.

much larger sample size than the mouth, had a flow rate of 32 mL/sec. Overbosch (*10*) noted that although in real mouth conditions, the Reynolds number is about 500, the headspace sampler by Lee (*11*) used gas flows with a Reynolds number on the order of 1. The Reynolds number for this simulator was about 14.

The stimulated saliva flow rate for individuals averages around 2 mL/min (*5*). Assuming a 5-g or 5-mL bite size and retention in the mouth for 30 seconds while chewing 1 mL of saliva secreted. Therefore, the sample size of 2.5 L required a stimulated saliva volume of 500 mL.

Oil and Water Retronasal Volatility. The collection of flavors on six Sep-Paks through the 15-minute experiment allowed an estimate of the dynamics of volatility. In looking at the liberation of the flavors (Figure 2), there was a definite difference between α–pinene in water and all of the other flavor compounds. α–Pinene in water volatilized almost immediately and leveled-off. 2-Methoxy-3-methyl pyrazine shown in Figure 2b is an example of the constant release over time exhibited by all of the compounds in the oil/water mixture and all but α–pinene in water. The curves can be analyzed in terms of first-order reaction kinetics:

$$\frac{d[A]}{dt} = -k[A] = -\left(\frac{d[A]}{dt}\right)_{\text{Sep-Pak}} \qquad (1)$$

where:
A = concentration of volatile in solution
k = first order rate constant.

The amount of collected on the Sep-Pak over time is what was actually measured rather than the amount remaining in solution over time. For all of the compounds but α–pinene in water, the amount volatilized is negligible compared to the amount remaining in solution. The following assumption can be made:

$$[A] = [A_o]$$

where A_o = initial concentration of volatile in solution.

Thus,

$$k = \frac{\left(d[A]/dt\right)_{\text{Sep-Pak}}}{[Ao]} \qquad (2)$$

The slopes of the graphs plotting µg flavor compound collected from the Sep-Pak versus time (Figure 2b) determined $(d[A]/dt)_{\text{Sep-Pak}}$. The rate constants, k (min^{-1}) were calculated using equation 2 for the compounds in the oil/water mixture and water (Table I). α–Pinene in water represented a case where the reaction (volatilization) proceeded to a much further stage. Its rate constant was estimated using the equilibrium first order rate equation:

$$\ln \frac{[A] - [A\infty]}{[Ao] - [A\infty]} = - kt \qquad (3)$$

where A_∞ = final equilibrium concentration of volatile in solution.

A_∞ was determined by the method of selected points where values of A_∞ were chosen so that a plot of ln $(A - A_\infty)$ versus time was linear.

Table I. Flavor Compound Volatilization Rate Constants, log P Values (P = Concentration in Octanol / Concentration in Water), and Estimated Vapor Pressures

Flavor Compound	*Ratio k water to k oil*	*water k (min $^{-1}$) x 10^{-5}*	*oil + saliva k (min $^{-1}$) x 10^{-5}*	*Log P*	*Vapor Pressure (mm), 25°C*
α–Pinene	7700	33000	4.3	3.75	4.7 [a]
Ethyl-2-methyl-butyrate	130	1400	11	1.19	8.8 [b]
1,8-Cineole	100	230	2.4	1.34	1.8 [a]
2-Methoxy-3-methyl-pyrazine	7	14	2.1	1.30	2.6 [c]
Methyl anthranilate	3	1.2	0.46	2.23	0.2 [d]
Vanillin	na	"0"	"0"	0.93	0.00005[d]
Butyric acid	1	62	64	0.79	1 [a]
Maltol	< 0	"0"	2.2	0.02	0.2 [c]

[a] Reported (*12*)

[b] Calculated with the Hass/Newton equation (*12*)

[c] Boiling point estimated by GC retention index and vapor pressure calculated with the Hass/Newton equation (*12*)

[d] Extrapolated from known vapor pressures above 25 °C

The rate constants show that there were large differences in the volatility of the flavor compounds. The variation of volatility between flavor compounds is very large in the water matrix but much smaller in the oil + "saliva" matrix. The highly volatile compounds all had large decreases in volatility with the addition of oil. Log P, a measure of polarity, was used to show differences in the rate constants' ratio between oil and water. The flavor compounds in this study, chosen for their range of polarity, did indeed span the range of log P values. Log P had been previously measured for butyric acid (0.79) and 2-methoxy-3-methyl-pyrazine (1.24) (*13, 14*). In this study, vapor pressure explains the low volatility of some compounds. Some of the values were not available, however, and the Hass/Neston equation (*15*) was used for estimation. For two of the compounds, boiling point information was not available so this was estimated by their retention on an OV-101 GC column.

Because α–pinene was practically insoluble in water, as would be expected by its high log P value, it had the largest volatility rate constant. Conversely, it was soluble in the oil matrix as seen by a decrease of 8000 in the volatility rate constant. The volatility of hydrocarbon flavors lacking functional groups, such as α–pinene, is highly dependent on the oil content of the food. Methyl anthranilate, also a highly non-polar compound, was quite non-volatile in both the water and oil matrices. This can be explained by its low vapor pressure in the pure state. Ethyl-2-methyl-butyrate had a 10-fold higher volatility than 1,8-cineole in both matrices but had similar partitioning in the water vs oil matrices. As would be expected by their nonpolarity, these two had a 100-fold higher volatility in water over oil. 2-Methoxy-3-methyl pyrazine had an intermediate volatility and did show a 7-fold oil solvation effect.

The more polar compounds, vanillin, maltol, and butyric acid did not show a decrease in volatility in the oil matrix. While butyric acid had a moderate volatility in each, vanillin was not detected in either, probably due to its low vapor pressure in the pure state. This may be why vanillin is frequently added at at least several hundred mg/L to foods. Vanillin, a potent compound in air with an odor threshold of 1.1×10^{-9} mg/L, shows a large reduction in threshold in water, 2.2×10^{-1} mg/L (*16*). Maltol, one of the most polar flavor compounds known, actually exhibited a larger volatility in oil than water. In water based foods, maltol needs to be added in large quantities for detection and is frequently found at 100 mg/L. Its threshold in water is indeed very high, 29 mg/L (*17*).

Predictive Parameters of Retronasal Flavor Volatility. Several properties of the flavors were evaluated for their prediction of relative flavor volatility using linear regression analysis. The first is vapor pressure over the pure compound at 25 °C. Vapor pressure was shown to relate to the air-solution partition coefficient of a homologous series of compounds (*18*) and to the relative volatiliation rates of pesticides (*19*). Although the compounds in this study represent a range of vapor pressures, they have very different structures and functional groups. As shown in Table II, the relationship between vapor pressure and volatility proceeds in the expected direction as larger vapor pressures result in a larger volatility in both matrices. However, the strength of the correlation (R^2) is low. Similarly, as the boiling point (which measures the same vapor pressure phenomena) increases, the volatility decreases for both matrices.

Log P, the partition coefficient between octanol and water, is widely used as a standard hydrophobicity parameter (*14*). Utilizing reported and calculated data, it was linearly related to the flavor threshold log ratio in oil and water (*20*). Significance was greatest with a homologous series of compounds. However, when the eight values were evaluated as predictors of relative oil or water volatility, a

Table II. Evaluation of Linear Regression of Predictors with the Volatility Rate Constant for the Two Matrices

Predictor (x)	*Slope and R^2 when y= log (volatility in water)*		*Slope and R^2 when y = log (volatility in oil + saliva)*	
Boiling Point (°C)	-0.03	$R^2 = 0.5$	-0.2	$R^2 = 0.5$
Vapor Pressure 25 °C (mm)	0.4	$R^2 = 0.5$	0.1	$R^2 = 0.1$
Log P	0.7	$R^2 = 0.2$	-0.1	$R^2 = 0.04$
Retention Time on OV-101 GC Column (min)	-0.2	$R^2 = 0.4$	-0.1	$R^2 = 0.8$

relation was not found, as seen by the low R^2. Similarly, when looking at several hundred compounds, a polarizability measure did not correlate with the air-water partition coefficient (*21*). Since Log P is a partition ratio between a nonpolar and polar phase, it is a better predictor of the volatility rate constant ratio between the two matrices. In general, a trend can be seen in Table I where the compounds with higher log P have a higher ratio, and the compounds with lower log P have reduced ratios.

Lastly, the retention times of the compounds on a methyl silicone (nonpolar) gas chromatography column were related to the volatility rate constants. The retention time is essentially a measure of the partitioning from the volatiles in helium to the nonpolar stationary phase. It is, consequently, not surprising that a good correlation was found with the volatility rate constants for the oil + saliva matrix and that a poor one was pbtained for the water matrix. The compounds with higher retention times have lower volatility.

Conclusions

The retronasal aroma simulator allowed the volatility measurement of aroma compounds with conditions similar to that found in the mouth. It was sensitive to flavor levels of mg/L. The volatility rate constants for the nonpolar flavor compounds were reduced in an oil as compared to a water matrix. Large differences were seen in the volatility of the different flavor compounds, and these relative differences were only roughly predicted by known chemical parameters. The measurement of the volatility in actual foods, especially those with reduced oil content, is a future application of the retronasal aroma simulator.

Acknowledgment

This material is based upon work supported under a National Science Foundation graduate fellowship.

Literature Cited

1. Voirol, E.; Daget, N. *Lebensm. Wiss. u. Technol.* **1989**, *22*, 399-405.
2. Voirol, E.; Daget, N. *Lebensm. Wiss. Technol.* **1986**, *19*, 316-319.
3. Kuo, Y.L.; Pangborn, R.M.; Noble, A.C. *Int. J. Food Sci and Tech.* **1993**, *28*, 127-137.
4. Roberts, D. D.; Acree, T, E. *J. Agric. Food Chem.* **1995**, *43*, 2179-2186.
5. Bourne, M.C. *In Food Science and Technology, a Series of Monographs,* Stewart, G. F.; Schweigert, B. S.; Hawthorn, J., Eds.; Academic Press: New York, 1982.
6. Shama, F.; Sherman, P. *J. Texture Stud.* **1973**, *4*, 111-118.
7. Cutler, A. N.; Morris, E. R.; Taylor L. J. *J. Texture Stud.* **1983**, *14*, 377-395.
8. Elejalde, C. C.; Kokini, J. L. *J. Texture Stud.* **1992**, *23*, 315-336.
9. Rao, M. A.; Cooley, H. J. *J. Texture Stud.* **1984**, *15*, 327-335.
10. Overbosch, P.;. Afterof, W. G. M.; Haring, P. G. M. *Food Reviews Int.* **1991**, *7*, 137-184.
11. Lee, W. E. I. *J. Food Sci.* **1986**, *51*, 249-250.
12. Weast, R. C., Ed. *Handbook of Chemistry and Physics.* 56th ed. CRC Press: Cleveland, OH, 1975.
13. Yamagami, C.; Takao, N.; Fujita, T. *J. Pharm. Sci.* **1991**, *80*, 772-777.
14. Leo, A.; Hansch, C.; Elkins, D. *Chem. Rev.* **1971**, *71*, 525-554.
15. Klein, R. G.. *Toxicology.* **1981**, *23*, 135-147.
16. Fazzalari, F.A., ed. *Compilation of Odor and Taste Threshold Values Data*; ASTM Data Series, ASTM: Philadelphia, 1978.
17. Bingham, A. F., et al., *Chem. Senses.* **1990**, *15*, 447-456.
18. Buttery, R. G., Ling, L. C.; Guadiagni, D. G. *J. Agric. Food Chem.* **1969**, *17*, 385-389.
19. Dobbs, A. J.; Hart, G. F.; Parsons, A. H. *Chemosphere* **1984**, *13*, 687-692.
20. Gardner, R. J. *J. Sci. Food Agric.* **1981**, *32*, 146-152.
21. Schuurmann, G.; Rothenbacher, *C. Fresenius Envir. Bull.* **1992**, *1*, 10-15.

Chapter 17

Investigation of the Interaction Between Dentifrice Flavor and Product Base by Principal Component Analysis of Headspace Gas Chromatography

John Brahms, James Masters, John Labows, and Michael Prencipe

Colgate-Palmolive Company, 909 River Road, Piscataway, NJ 08855–1343

The extent to which dentifrice flavor interacts with the product's continuous phase and the abrasive profoundly affects how flavor will be perceived. Static headspace-GC analysis was used to examine dentifrice in the neat form and as 50% and 25% (v/v) suspensions in water to simulate the effect of dilution upon brushing. To aid in our investigations of the vapor-liquid equilibria of flavor, data were examined by principal component analysis. Two abrasives and two continuous phase compositions were considered. The two abrasives differ in their ability to complex flavor thus affecting the headspace profile of the neat dentifrice. However, flavor release upon dilution is related to continuous phase composition and is independent of the abrasive composition.

A toothpaste consists primarily of an abrasive and a liquid continuous phase. The major function of the abrasive is cleaning. Several abrasives employed include silica, alumina, dicalcium phosphate dihydrate, insoluble metaphosphate or calcium pyrophosphate. Depending on the chemical nature of the abrasive, it can interact weakly or strongly with flavor oils, which, in turn, impacts the vapor phase concentration of flavor components in equilibrium with the solid and liquid phase. The liquid continuous phase consists of water, salts, humectant, surfactant, polymeric rheology modifying agents, and flavor oil. An important factor influencing perceived flavor intensity is the effective vapor pressure of the flavor in the system. For this reason, headspace-GC has proven to be a valuable tool in investigating the interactions of flavor and non-flavor components in a variety of product matrices. A number of studies have been conducted to determine how these interactions impact flavor perception *(1-3)*.

In a system that contains only surfactant, flavor and water, the flavor will partition between the surfactant and the continuous pseudo-phases. The chemical

0097–6156/96/0633–0188$15.00/0

structure of the individual flavor compounds will have a profound effect on how this partitioning takes place. The ratio of flavor to surfactant also has a strong effect on the partitioning behavior of flavor between these two phases. At low flavor/surfactant ratio, a flavor compound can be thought of as an infinitely dilute solute partitioning between an aqueous and an organic phase. Under these conditions, micelle/continuous phase partition coefficients can be predicted with a fair degree of accuracy by the log P value of the flavor material *(4-5)*. This relationship is not as straight forward, however, when comparing compounds over a wide range of functionality or at the high flavor/surfactant ratios that exist in toothpaste. At high flavor/surfactant ratios, a flavor partitioning transition is observed between free surfactant and water during emulsion formation. Addition of other components, such as thickeners and humectants, results in even more complex behavior.

At high flavor/surfactant ratios, the surfactant forms an emulsion with flavor *(6)*. The surfactant cannot be treated as an amorphous organic phase. The structure of the solute as well as the concentration of solute and surfactant at the phase boundary play an important role in solubilization *(7)*. Theoretical models for the effect of solute molecular structure on solubilization behavior under these conditions have been developed *(8)*. Molecules that are rigid hydrophobic structures reside in the core region of the micelle. This is called Type-I solubilization. Solute molecules that look like surfactants will act as co-surfactants, orienting themselves so that they will have favorable interactions with both head and tail protions of the surfactants in the micelle structure. This is called Type-II solubilization.

A dentifrice continuous phase is not merely a surfactant-flavor-water system since it also contains humectant as well as viscosity modifying agents. Surface tension measurements have shown that sorbitol has the effect of promoting micelle formation thus reducing the critical micelle concentration (CMC). Glycerin on the other hand, increases the CMC, interfering with the formation of micelles *(9)*. Viscosity modifiers don't have as great an effect on equilibrium release but do have a strong effect on the kinetics of release *(2)*.

Previous work has shown that dentifrice ingredients can interact with flavor components impacting how a formulation's taste will be perceived *(1-2)*. These studies have treated flavor as a single entity, concentrating on the effect of non-flavor ingredients on the sum of the headspace levels of the individual flavor components. This approach has proven useful in providing an understanding of the relative importance of non-flavor ingredients as well as the relative importance of kinetic and thermodynamic effects on flavor release caused by dilution which occurs during brushing. In order to understand how changes in product formulation affect flavor character, it is necessary to understand the effect of such changes on individual flavor components.

In this chapter we will discuss experiments that have been carried out to more fully understand the physical and chemical interaction between individual dentifrice flavor components and product base ingredients. Headspace-gas chromatographic analysis has been used to investigate the interaction of flavor components and other dentifrice ingredients. Our ultimate objective was to correlate available flavor to perceived intensity and character, as well as provide insights into the factors that affect flavor availability. To this end, headspace-GC was employed to (i) determine the interaction between the abrasive and individual flavor components and (ii) investigate what effect continuous phase composition has on the release profile of the flavor over a range of dilutions expected to occur during brushing. The latter is complicated by the unique release behavior exhibited by different flavor components. To aid in our analysis of the data, principal component analysis was used to identify how various flavor compounds are affected by the two types of formula modification.

A GC profile can be represented as a point in a an *n*-dimensional space where *n* is equal to the number of volatile compounds of interest and the level of each component is the magnitude along a given axis. The relative similarity or difference between GC profiles can be represented by the distances between these points. It is impossible, however, to visualize in more than two or three dimensions.

Principal component analysis is a technique whereby the original variables are converted to a new set of variables called components, each of which is a linear combination of original variables. This is achieved by performing a rotational transformation on the original data. Before this is done, the original variables are usually scaled to eliminate biases such as GC response factors and order-of-magnitude differences in the mean levels of different volatile compounds. The components which account for a disproportionate amount of variance are called principal components. The coefficients for the (scaled) original variables are called loadings. The first principal component (PC-1) accounts for the largest percentage of the variance followed by the second principal component (PC-2). By plotting the scaled original data on an x-y graph consisting of the first two principal components, a two dimensional representation of the original data is generated in which most of the information content is preserved *(10)*.

PCA has proven useful as a classification tool for comparing complex GC profiles *(10-13)*. Further, this technique has been used extensively in flavor and fragrance research to reduce the dimensionality of complex data *(10,14)*. In this study, PCA was employed to differentiate the interactions of flavor components with dentifrice continuous phase from flavor-abrasive interactions. By analyzing individual flavor components for a set of different product formulations at several dilutions, it was possible to determine the effect of variations in product formulation on both neat dentifrice headspace flavor profiles and flavor release during brushing. These results also provided insights into how modification of continuous phase composition and abrasive affect the availability and release of individual flavor components. We found that interactions of flavor with non-flavor ingredients differ significantly for different flavor components as a function of dilution.

Experimental

Design of Experiment. In this study, two variables were considered: abrasive composition and continuous phase water/humectant ratio. The two abrasives investigated were amorphous silica and dicalcium phosphate. The effect of continuous phase composition was investigated by varying the water/humectant ratio in the continuous phase. Two continuous phase compositions were investigated, one with a high water/humectant ratio and one with a low water/humectant ratio. A total of three product formulations were used:

Formula 1: Silica abrasive containing high water/humecant ratio
Formula 2: Silica abrasive containing low water/humectant ratio
Formula 3: Dicalcium phosphate abrasive containing high water/humectant ratio.

The formula containing dicalcium phosphate abrasive and low water/humectant ratio was not available for inclusion in this study. Although, inclusion of the fourth formula would have better enabled us to detect mixture effects of the two variables, available evidence suggests that the effect of abrasive on neat dentifrice flavor profile and the effect of continuous phase composition on flavor release are independent.

During brushing, the dentifrice is diluted by about 3–4 fold with saliva *(15)*. In order to investigate the effect of dilution on headspace flavor profile, samples of dentifrice were examined by static headspace-GC analysis at full strength and at 50 % (v/v) and 25% (v/v) suspensions in water. The same sample volume for each dilution was used, maintaining a constant ratio between liquid and gas phase volumes in the headspace vials for all samples.

The effects of the three formula compositions on fourteen major flavor components were visualized using principal component analysis. Three different formulations at three dilutions, gave a total of nine samples yielding an under determined data set for principal component analysis. Two additional formulas, prepared and aged at the same time as formulas 1–3 were included in the original data set for this experiment. The two additional formulas were compositionally identical to formulas 2 and 3 except that the method of preparation was modified. These formulas will be referred to as follows:

Formula 2a: Same composition as Formula 2, different preparation method
Formula 3a: Same composition as Formula 3, different preparation method.

The change in preparation method had little effect on neither neat dentifrice headspace profile nor flavor release behavior. The preparation method will not be considered as a variable in the experiment.

Sample Preparation. Samples of dentifrice consisting of abrasive, sodium lauryl sulfate (SLS), humectant, water, thickener and flavor were prepared on the same day and aged for six weeks at room temperature in glass jars protected from light. All dentifrices used in this study contained the identical flavor and surfactant levels. After aging, samples for headspace-GC analysis were prepared and analyzed. Neat dentifrice headspace-GC samples were prepared by placing 5.00 mL of product into a 22 mL headspace-GC vial which was then sealed by an aluminum lined septum cap. The 50% suspensions were prepared by first placing 2.50 mL of dentifrice into 22 mL headspace-GC vials followed by 2.50 mL of deionized water. The water and dentifrice were stirred with a glass rod for a length of time sufficient to yield an homogeneous suspension, then sealed. The 25% suspensions were prepared by the same procedure as the 50% suspension samples except that 1.25 mL of dentifrice and 3.75 mL of deionized water were used.

In order to investigate the flavor release behavior of the dentifrice continuous phases in the absence of abrasive, the continuous phases were prepared without thickener or abrasive. From these continuous phase solutions, a series of dilutions were performed starting at full strength and successively diluting with equal volumes over eleven dilutions. Samples of each dilution (5 mL) were placed in 22 mL headspace-GC vials and examined under conditions identical to those used for the dentifrice samples.

Instrumental Analysis. The samples were examined by headspace-GC analysis using a Perkin-Elmer Sigma 2000 Capillary Gas Chromatograph equipped with an HS-100 automated headspace injection system and a flame ionization detector. A 30 m x 0.3 mm fused silica Carbowax capillary column was used for all sample analyses. The vials were thermostatted at 60 °C for 1 hour and pressurized with nitrogen for 1 minute. The headspace vapor was then introduced into the gas chromatograph. GC conditions were: column held at 50 °C for 2 min., ramped to 175 °C at 6 °C/min., held at 175 °C for 2 min. The data were collected via a Nelson-900 series interface connected to a Gateway 386/33 computer using the PE-Nelson Turbochrom-3 GC data acquisition and analysis software. GC peak areas for the major flavor components were tabulated and exported to a comma-delimited text file.

Results

Reproducibility of Headspace-GC Data. Five replicates of each sample at each dilution were analyzed. Coefficients of variance were found to be independent of product formulation. However, the neat dentifrices exhibit a higher degree of variability than either the 50% or 25% samples. This is attributed to the GC sampling conditions. The neat dentifrice samples were highly viscous and required a longer time to reach equilibrium at the headspace-GC sampling conditions than either the 50% (v/v) or 25% (v/v) samples. Increasing equilibration time did not significantly change the headspace levels of the neat samples but did reduce variability slightly. However, longer equilibration times result in flavor degradation for the diluted samples. One hour equilibration proved the optimum balance between sample stability and reproducibility. Coefficients of variance for the 14 flavor components based on five replicate samples at each concentration are summarized in Table I.

Effect of Abrasive on Individual Flavor Components. In order to determine the effect of abrasive on individual fragrance components, headspace-GC of neat products containing each abrasive were compared. The relative difference in headspace-GC peak areas between silica and dicalcium phosphate containing dentifrices was determined for each of the fourteen fragrance components. T-tests were performed to determine whether the headspace-GC peak areas were

Table I. Coefficients of Sample Variance for Major Flavor Components

	A	B	C	D	E	F	G
Full Strength	9.98%	9.15%	13.96%	16.55%	3.80%	3.88%	27.89%
50% (v/v)	3.03%	2.08%	4.89%	4.44%	1.32%	1.48%	4.21%
25% (v/v)	4.09%	6.03%	8.77%	9.97%	3.64%	1.44%	4.44%

	H	I	J	K	L	M	N
Full Strength	3.54%	6.98%	5.18%	2.37%	1.68%	2.36%	1.88%
50% (v/v)	1.73%	5.43%	13.15%	1.80%	1.27%	2.32%	2.19%
25% (v/v)	3.03%	2.81%	4.23%	1.51%	1.05%	3.28%	2.23%

significantly different for samples containing silica vs. dicalcium phosphate abrasive. All samples contained the same continuous phase composition. Results are summarized in Table II. The neat dentifrice headspace concentrations for all but three flavor components are significantly different for silica vs. dicalcium phosphate containing dental creams.

Release Behavior of Continuous Phase. The differences in release behavior between the high and low water formulas, can be seen by comparing the effect of dilution on the equilibrium headspace concentration of each individual flavor component. Comparisons between the mean headspace concentrations of each flavor component for full strength and the 50% (v/v) suspensions, and between the

Table II. Peak Areas of Major Flavor Components for Silica Relative to Dicalcium Phosphate-Neat Dentifrice

Flavor Component	Rel. Peak Area	p-Value
A	same	NS
B	less	0.004
C	less	0.005
D	less	0.004
E	less	<0.001
F	less	0.004
G	less	NS
H	greater	<0.001
I	same	NS
J	same	NS
K	less	<0.001
L	less	<0.001
M	less	0.006
N	less	<0.001

Table III. Flavor Release on Dilution for High Water/Humectant Formulas

	Full Strength → 50% (v/v)		50% → 25% (v/v)	
Flavor Comp.	Change	p-Value	Change	p-Value
A	increase	<0.001	no change	NS
B	no change	NS	decrease	0.0038
C	increase	<0.001	no change	NS
D	increase	<0.001	no change	NS
E	no change	NS	decrease	0.0016
F	increase	<0.001	no change	NS
G	increase	<0.001	increase	NS
H	no change	NS	no change	NS
I	increase	<0.001	increase	<0.001
J	increase	<0.001	increase	<0.001
K	increase	<0.001	increase	0.0014
L	increase	<0.001	no change	NS
M	no change	NS	decrease	<0.001
N	increase	<0.001	increase	<0.001

Table IV. Flavor Release on Dilution for Low Water/Humectant Formulas

	Full Strength → 50% (v/v)		50% → 25% (v/v)	
Flavor Comp.	Change	p-Value	Change	p-Value
A	increase	<0.001	increase	<0.001
B	increase	<0.001	no change	NS
C	increase	<0.001	increase	<0.001
D	increase	<0.001	increase	<0.001
E	increase	<0.001	no change	NS
F	increase	<0.001	increase	<0.001
G	no change	NS	increase	<0.001
H	increase	<0.001	no change	NS
I	increase	0.0053	increase	<0.001
J	increase	NS	increase	0.0062
K	increase	<0.001	increase	<0.001
L	increase	<0.001	no change	NS
M	no change	NS	decrease	<0.001
N	increase	<0.001	increase	<0.001

50% (v/v) suspensions and the 25% (v/v) suspensions are summarized in Tables III and IV. For the complete dentifrice formulas containing a high water/humectant ratio in the liquid phase, there is a statistically significant increase in headspace flavor level for all but four flavor components upon dilution from full strength to 50% (v/v). In the formulas containing a low water/humectant ratio, all flavor components except compounds G and M increased upon dilution from full strength to 50%. No flavor components exhibit a significant decrease on the first dilution. The high and low water/humectant continuous phase formulas exhibited different behavior when comparisons were made between the 50% and 25% (v/v) suspensions. For the high water/humectant formulas, only five of the flavor components show an increase upon dilution while the rest either remain unchanged or decrease. The low water continuous phase dentifrice formulas, however, show an increase for nine flavor components going from 50% to 25% (v/v) concentration.

The observation that some flavor components increase while others decrease or remain relatively unchanged can be understood if we consider the effect of dilution of the liquid phases in the absence of abrasive. A series of samples was prepared in order to understand how dilution affects the equilibrium headspace concentration of flavor. The undiluted sample consisted of a dentifrice continuous phase which contained 0.9% of the flavor mixture. A second sample was prepared by diluting this mixture with an equal volume of water. A portion of the diluted solution was treated likewise and the procedure was repeated until a series of 12 dilutions were prepared. Each dilution contained one half of the concentration of continuous phase in the previous sample. In Figure 1, dilution factor indicates 2^{-n} multiplied by the concentration of the original solution. The Y-axis shows headspace concentration normalized to the level for the full strength sample.

As can be seen from Figure 1, dilution of the dentifrice continuous phase results in an increase in total GC peak area for the first four dilutions. As the mixture consisting of a fixed ratio of flavor to surfactant is diluted, the water/surfactant ratio gradually increases and hence the amount of surfactant available for solubilization of flavor decreases. As the concentration of surfactant micelles decreases, the solubility of flavor in the solution declines with a corresponding increase in headspace flavor level. This lowered solubility more than offsets the reduction of flavor concentration due to dilution.

Figure 2 shows the effect of dilution on three individual flavor components. The headspace-GC peak for each component is normalized to headspace level of the undiluted sample. The headspace level of flavor with high micelle solubility actually increases over several dilutions. Compound A in Figure 2 shows this type of behavior. Highly water-soluble flavor molecules or those with low micelle solubility are not as strongly affected by the reduction in micelle phase. These compounds decrease upon dilution over the entire concentration range. Compound C in Figure 2 shows this type of behavior. Compound B exhibits intermediate behavior.

Principal Component Analysis. Principal component analysis was used to investigate the effect of abrasive and continuous phase simultaneously, as they affect different flavor components. The five formulae were examined by headspace-GC at full strength, as well as 50% and 25% (v/v) suspensions in deionized water. The tabulated GC peak areas for 14 major flavor components for the 15 samples were

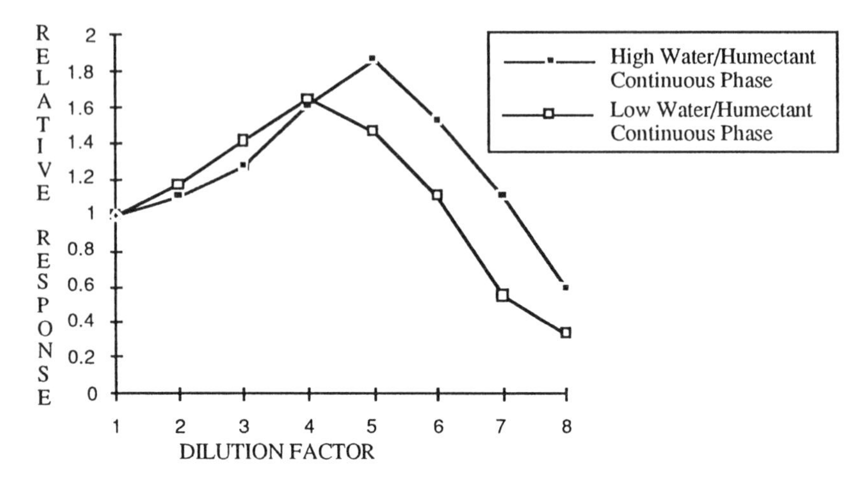

Figure 1. Headspace flavor level of continuous phase as a function of dilution.

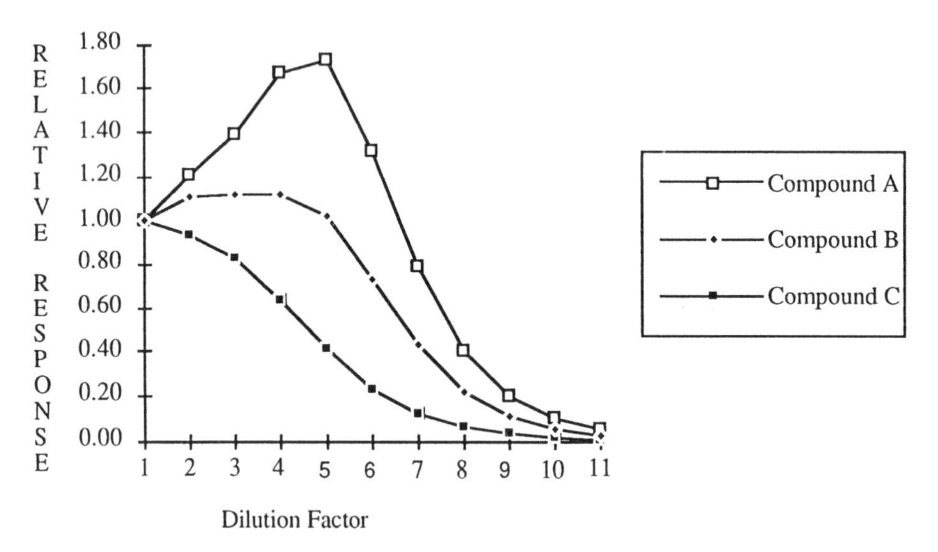

Figure 2. Release behavior of different flavor components.

autoscaled in order to eliminate biases due to the different levels and GC response factors of different flavor compounds.

Principal component analysis was performed on the autoscaled GC data set using the EinSight pattern recognition program (Infometrix, Inc. 2200 6th St, Suite 833, Seattle Wa, 98121, Copyright 1989). The first two principal component eigen vectors account for nearly 83% of the sample variance. The loadings for the first two principal components are shown on Figure 3. Numerical values are listed in Table V. Each of the flavor materials are indicated by the letters A through N.

Table V. Loadings for PC-1 and PC-2

	A	B	C	D	E	F	G
PC-1	0.271	0.204	0.242	0.316	0.052	0.321	0.315
PC-2	0.360	0.045	0.345	0.049	0.517	0.024	0.045

	H	I	J	K	L	M	N
PC-1	0.080	0.311	0.297	0.264	0.311	-0.255	0.313
PC-2	-0.425	-0.153	-0.118	-0.161	-0.226	0.303	-0.007

To gain a better understanding of the dilution behavior, the changes in GC profiles as a function of dilution were investigated. The scores for the 15 samples are shown plotted against the first 2 principal components in Figure 4. It can be seen that the dental creams formulated with silica as the abrasive form a tight cluster with the dicalcium phosphate sample as an outlying data point. As mentioned earlier, for samples with silica and different continuous phase compositions, the neat dentifrice headspace profiles were virtually identical for most of the flavor components.

Upon dilution, two of the dental creams formulated with silica as an abrasive exhibited continued progress in the positive-x direction when diluted to 50% and to 25% with water. For the other two, there appears to be a fall-off after the first dilution. By connecting the points, a map of the release profile for the flavor from a particular formulation upon dilution can be generated as shown on Figure 5.

Interpretation of Principal Component Loadings. As shown above, the headspace concentrations of all flavor components increase on dilution from full strength to 50%. Since most flavor components have positive loadings for the first principal component, changes that occur upon dilution are reflected as changes in the first principal component score. The main difference between the samples that contained the different continuous phases was that which occurs between the second and third dilution. The samples that contained low water/humectant continuous phase showed a greater increase in the positive x direction than those that contained high water/humectant continuous phase. Changes in the first principal component

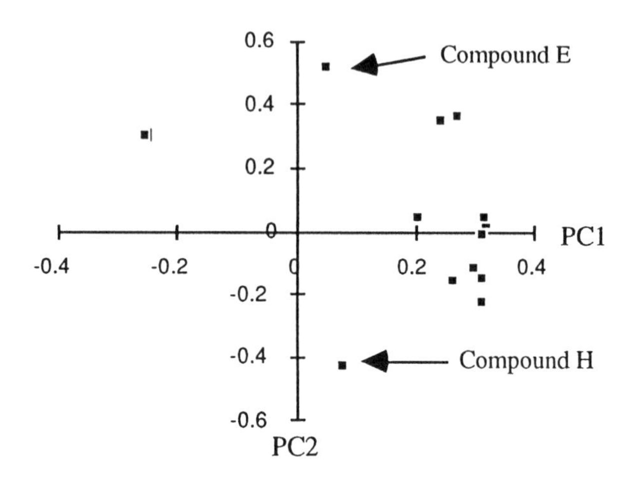

Figure 3. Principal component loadings plot.

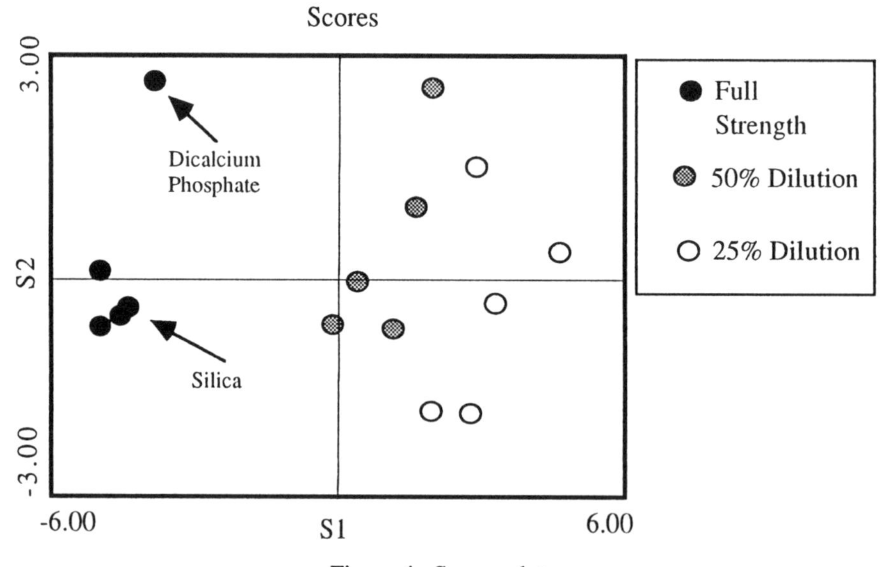

Figure 4. Scores plot.

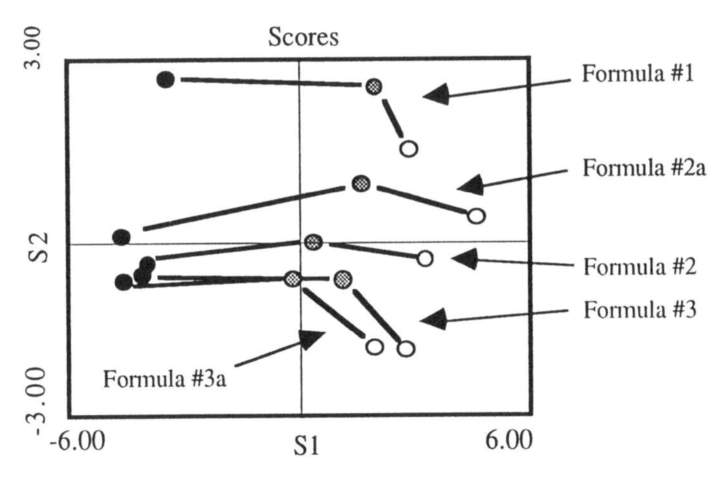

Figure 5. Map of dilution profiles.

score give a visual representation of the effect of dilution on the neat dentifrices. The first component can therefore be named as being related to dilution effects.

Earlier it was shown that one of the flavor components, compound H, is the only material that has a greater GC peak area in the silica-containing formula than in the dicalcium phosphate formula. Compounds E and H have high loadings for the second principal component and low loadings for the first principal component. Examination of the loadings plot provides some interesting insights into the nature of the interaction of flavor with abrasive. Two flavor components, having large positive and negative loadings for the second principal component were found to chemically interact with the silica abrasive. This would indicate that the second principal component is essentially related to the relative effect of the two abrasives on neat dentifrice headspace.

Conclusion

Dentifrice is a complex multi-component multi-phase system not unlike many food products. Flavor can interact with both the liquid and solid phases of the product. It has been shown that the release behavior which occurs with dilution depends on both the continuous phase composition and the chemical nature of the flavor ingredients. Flavor materials with higher micelle solubility will have a greater release upon dilution. The micelle solubility of a fragrance material depends upon a number of factors, such as its water solubility, as well as its ability to undergo thermodynamically favorable interactions with both the head group and tail portions of surfactants present in the micelle. Since different flavor compounds undergo different release behavior, they must be considered separately in order to fully understand the effect of any potential modifications in base composition on the flavor release behavior of the final product.

Principal Component Analysis can be used to identify data patterns which are not obvious by other techniques. In this study it was shown that the headspace profile for the a dental cream formulated with one type of abrasive was significantly different from that containing a different abrasive. The formulas containing silica and a high water/humectant continuous phase show release behavior different from the formulas containing the same abrasive and a low water/humectant continuous phase. In order to gain a complete picture of flavor/non-flavor interactions, it is necessary to look at the behavior of a variety of components simultaneously and not treat flavor as a single entity. Principal components analysis identifies which flavor materials are most affected by change in abrasive or continuous phase. The first two principal components are qualitatively related to different effects. The first component was found to be correlated with release of flavor upon dilution., while the second was related to differences in headspace flavor profile due to interaction of the different abrasives with other flavor components.

Literature Cited

1. Overbosch, P.; Afterof, W. G. M.; Haring, P. G. M. *Food Reviews International* **1991**, *7*, 137-184.
2. Robinson, R. S.; Tavss, E. A.; Santalucia, J.; Carroll, D. L. *J. Chromatog.* **1988**, *455*, 143-149.
3. Robinson, R. S.; Tavss, E. A.; Santalucia, J. *J. Soc. Cosmet. Chem.* **1988**, *39*, 305-314.
4. Pim, G. H. M.; Muijselaar, H. A.; Claessens, C. A.; Cramers *Anal. Chem.* **1994**, *66*, 635-644.
5. Valsaraj, K. T.; Thibodeaux, L. J. *Separation Sci. and Tech.* **1990**, *25*, 369-395.
6. Tokuoka, Y.; Uchiyama, H.; Abe, M.; Ogino, K. *J. Coll. Int. Sci.* **1992**, *152*, 402-409.
7. Bourrel, M.; Biais, J.; Bothorel, P.; Clin, B.; Lalanne, P. *J. Disp. Sci. and Tech.*, **1991**, *12*, 531-545.
8. Nagarajan, R.; Ruckenstein, E. *Langmuir* **1991**, *7*, 2934-2969.
9. Labows, J. N. *JAOCS* **1992**, *69*, 34-38.
10. Chien, M.; Peppard, T.; *IFT Basic Symp. Ser.* **1993**, *8*, 1-35.
11. Saxberg, B. E. H.; Duewer, D. L.; Booker, J. L. *Anal. Chem. Acta.* **1978**, *103*, 210-12.
12. Marengo, E.; Baiocchi, C.; Gennaro, M. C.; Bertolo, P. L. *Chemometrics and Intelligent Laboratorie Systems* **1991**, *11*, 75-88.
13. Aishima, T.; Nakai, S.; *J. Food Sci.* **1987**, *52*, 939-42.
14. Brahms, J. C.; Ziemann, B. M.; Labows. J. N. *Classification of Product Variants by Principal Component Analysis of Headspace-GC Data*; 31st Eastern Analytical Symposium, Somerset, N.J., November 16-18, 1992.
15. Burke, M. R.; Bertino, M.; Brahms, J. C.; Prencipe, M. *Surfactant Behavior in Dentifrice Formulations;* 83rd American Oil Chemists Society Meeting, Toronto, May 10-14, 1992.

Complex System Approaches

Chapter 18

Flavor Evaluation of Cheddar Cheese

Conor M. Delahunty[1], John R. Piggott, John M. Conner, and Alistair Paterson

Food Science Laboratories, Department of Bioscience and Biotechnology, University of Strathclyde, 131 Albion Street, Glasgow G1 1SD, United Kingdom

Cheese flavor is an articulated behaviorial response by consumers to a cheese-induced stimulus. This stimulus is induced by a balance of odorants and tastants released during cheese consumption. *In vitro* methods of cheese analysis measure what is present in a cheese, but they cannot determine which specific compounds are responsible for cheese flavor. To do so, consumers must be involved. Sensory evaluation methods are the scientific means of doing this. To date instrumental and sensory methods have, by and large, remained separate. The relationship between the two is sought through statistical correlations, the result being a rather loose and uncertain relationship based more on association than cause. One reason for this arises from the physicochemical diversity of those compounds which may be responsible for flavor, thereby affecting their release behavior during consumption. We have developed a method of analyzing volatile compounds released from cheese during its consumption, thereby eliminating a large part of this variability. We shall build on this approach to improve the understanding of Cheddar cheese flavor.

The study of Cheddar cheese flavor has been ongoing for over fifty years. Much has been accomplished, yet most would agree, the understanding remains inadequate. What flavor compound balance is required for a good cheese flavor? This question remains unanswered. Reasons for this can be attributed to the complexity of the cheese itself and to the complexity of consumers, who perceive and express flavor.

Cheddar cheese is a complex food system, consisting of moisture, milk fat, whey proteins, sugars, minerals, vitamins, and microflora embedded or partially embedded in a casein matrix. Within this are found other minor compounds, some of which contribute to flavor. The volatiles alone, thought to be the main contributors to cheese flavor quality, include such compounds as fatty acids, methyl, ethyl and higher

[1]Current address: Department of Nutrition, University College, Cork, Ireland

0097–6156/96/0633–0202$15.00/0

esters, methyl ketones, various aliphatic and aromatic hydrocarbons, short- and long-chain alcohols, aromatic alcohols, aldehydes, amines, amides, phenols and sulfur compounds (*1*). The physical and chemical properties of these compounds are different, and this will affect their release from the cheese structure and subsequent perception.

Cheese flavor is a consumer's expressed behaviorial response to a cheese-induced stimulus, i.e. interaction of the human senses, through olfaction and gustation, with the numerous chemical components of cheese (*2*). Therefore, for flavor expression, the cheese must be taken into the mouth and consumed, when addition of saliva coupled with mastication processes initiate the release of a flavor compound mixture. Unfortunately from the flavor scientist's point of view, consumers vary in their response. Flavor perception, and communication of flavor perception, will differ between individuals as a result of their physiological and sociological differences (*3,4*),although there are broad similarities. Most could easily identify a particular cheese and distinguish it from another variety.

Considering the complexity of cheese flavor, the means of cheese flavor evaluation, both instrumental and sensory, must be tackled intelligently and as simply as possible. A current concern is reduced-fat cheeses, developed in response to changing consumer attitudes to diet and health. Many recent products do not meet the flavor quality requirements of consumers (*5,6*). Improving flavor is difficult unless one understands how cheese flavor is formed, and which compounds give rise to cheese flavor perception.

The aim of the work described in this chapter was to develop methodology to establish what components, and in what balance, are responsible for Cheddar cheese flavor. We have developed a method of analyzing volatiles released from the cheese during mastication and consumption. By coupling this methodology with existing instrumental and sensory methods, we will identify some of the components responsible for Cheddar cheese flavor and assess their importance.

Experimental

Chemical Analyses. Twelve Cheddar cheeses, six traditional and six reduced in fat content, were purchased locally. In duplicate, volatiles from each cheese were analyzed by conventional, *in vitro*, headspace methodology. A grated 50-g cheese sample was allowed to equilibrate in a 1-L sealed flask held at 37 °C for 30 min and then flushed, for 30 min, onto a trap containing 150 mg Tenax-TA (60-80 Mesh) (Phase Separations Ltd., Deeside CH5 2NU, UK) with 60 ± 10 ml min^{-1} purified nitrogen.

For comparison, in duplicate, volatiles released during consumption of the cheeses were analyzed by *in vivo* headspace methodology developed for this purpose (*7*). Cheese (50 g) was consumed over a period of 30 min. Headspace released during consumption was displaced from the buccal cavity, via the nose, with a vacuum pump (Speedivac – High Vacuum Pump, Associated Electrical Industries Ltd., Newcastle, Staffs., UK) at a rate of 60 ± 10 mL min^{-1} and concentrated on the same Tenax-TA precolumn as the conventional methodology. A buccal headspace blank was taken each morning and after lunch. Blanks were also taken to determine volatiles remaining in the mouth after cheese consumption. A blank consisted of 30 min trapped buccal headspace without cheese consumption.

Traps were eluted with 2 mL diethyl ether (BDH Laboratory Supplies, Merck Ltd., Poole, Dorset BH15 1TD, UK) containing vanillin (Aldrich Chemical Co. Ltd., Gillingham, Dorset SP8 4JL, UK) at 10 μg mL^{-1} as internal standard (to correct for concentration and injection volume errors). The ether was evaporated to 10 μL with purified nitrogen and 1 μL injected via the septum programmable injector (230 °C) of a Finnigan-MAT ITS-40 gas chromatograph-mass spectrometer (Finnigan-MAT, Hemel Hempstead, Herts HP2 4TG, UK). The column was a 30 m x 0.25 mm Carbowax BP20 (df = 0.25 μm) (SGE (UK) Ltd., Milton Keynes MK11 3LA, UK), programmed from 40 °C (3 min) to 100 °C at 4°C min^{-1} then to 220 °C at 6 °C min^{-1}.

Analyses of variance (ANOVA) of chemical data were carried out using the GLM command of Minitab release 8. Data matrices were analyzed by principal component analysis (PCA) (*8*).

Assessment by Sniffport. Sniffport assessment of odorous volatiles was carried out for a mature Scottish Cheddar purchased locally. The approach, as for earlier experimental work, was to compare conventional headspace methodology with buccal headspace methodology. Duplicate 1000-g samples were used for each analysis. One "sample extract" for sniffport assessment consisted of 20 x 50 g sample extracts combined and reduced with purified nitrogen. Methodology for conventional headspace was as previously used. For *in vivo* methodology each 50-g sample of cheese was sub-divided into 10 x 5-g "bite-sized" cheese pieces. The entire sample was consumed, in a natural fashion, over a 5-min period, i.e. 30 sec per cheese piece. A total of 200 g of cheese was consumed each day for five days, allowing at least 1 hour between samples. The methodology was otherwise the same as previously used. The sniffport (SGE (UK) Ltd., Milton Keynes MK11 3LA, UK) was fitted to a Carlo-Erba 5300 chromatograph (Fisons Instruments Ltd., Crawley RH10 2QQ, UK) containing a Carbowax column (12 m x 0.53 mm Carbowax BP20 (df = 1 μm)(SGE (UK) Ltd., Milton Keynes MK11 3LA, UK), programmed from 40 °C (5 min) to 180 °C (2 min) at 6.1 °C min^{-1}. Three assessors, having previous experience of assessing column effluent from a sniffport, assessed the emerging odorous volatiles. Aromagrams were recorded on the time-intensity (T-I) module of the PSA-System (Oliemans, Punter & Partners, PO Box 14167, 3508 SG Utrecht, The Netherlands) set to collect 1 data point sec^{-1}. Odor descriptions were recorded on tape recorder. Peaks recognized only once were deleted. Relative peak intensities were finalized by discussion. Emphasis was on qualitative rather than quantitative assessment.

Descriptive Sensory Analysis. A vocabulary specific for Cheddar cheese flavor was generated, as follows, from a vocabulary previously developed for Cheddar cheese assessment (flavor and texture) in this laboratory (*9*). Twenty Cheddar cheeses were purchased locally. These were assessed, in duplicate, by a trained panel of 15 assessors. Samples were stored at refrigerator temperature until 15 min before sampling, and then allowed to warm to room temperature. Samples were presented as 5-g cubes in glasses, covered with watchglasses, and assessed in individual sensory booths under red lighting to minimize color differences. Assessors rinsed with deionized water between samples. Eight samples, in randomized order, were presented per session. Data were collected using the PSA-System. The panellists were invited to suggest extra terms as they felt necessary. After statistical analysis of the data (PCA, generalized Procrustes analysis (GPA; *10*), ANOVA) a revised vocabulary (which contained 14 descriptors) was selected which described or differentiated the cheeses tested. Subsequently, eight mild cheeses (four full-fat and four reduced-fat) selected to be as alike as possible in all other respects, were assessed using this vocabulary. Statistical analysis (PCA, GPA, ANOVA) revealed those descriptors which were most discriminating and these were selected to be used for T-I scaling. Descriptive analysis was used in this case not to distinguish between cheeses, but to select descriptors for temporal sensory assessment.

Time-intensity Scaling. Four mild cheeses, two full-fat and two reduced-fat, an arbitrary subset selected at time of purchase of the 8 assessed by descriptive analysis, were assessed, in duplicate, by a panel of 10 (of the 15) assessors for intensity of the descriptors cheesy and fruity for a time period of 90 sec. The samples were presented in the same manner as for descriptive analysis, except that only four samples were presented per session. Data were collected on the T-I module of the PSA-System. The relationships between descriptor and fat content and between descriptors relative to each other, with time of consumption, was sought. To view these, curves were averaged for each descriptor for each fat-content.

Results

Comparative Headspace. Twenty-nine compounds, none of which were detected in substantial quantities in the blanks, were chosen from the chromatograms as representative of all cheeses. Figure 1 shows a plot of the logarithm of the peak areas of these compounds for two cheeses, a mature Scottish Cheddar and a mature reduced-fat Cheddar type cheese, by both methods of analysis. Clear differences can be seen, between methods and between cheeses. Similar differences between methods were found for the other cheeses, and between the full-fat and reduced-fat cheeses. The pooled standard deviations for all compounds, calculated from the replicate analysis are shown in Figure 1. Analyses of variance showed significant differences ($p < 0.01$) between methods, and between fat contents, for several of these peaks, and a significant interaction ($p < 0.01$) between headspace method used and the fat content of the cheeses for several others (*7*).

A PCA of the data illustrates the relationships among the data sets, the volatile components released by the two differing methodologies, and those from the cheeses of differing fat content. Sample scores of the replicate analyses (the first two components accounted for 24% and 14% respectively) are shown in Figure 2. One-way analysis of variance showed significant discrimination between samples on both components.

Sniffport. Aromagrams for the *in vitro* (conventional headspace) and *in vivo* (buccal headspace) analysis are shown in Figures 3 and 4. In general the *in vitro* methodology produced a more concentrated extract with odorous compounds not found in that of

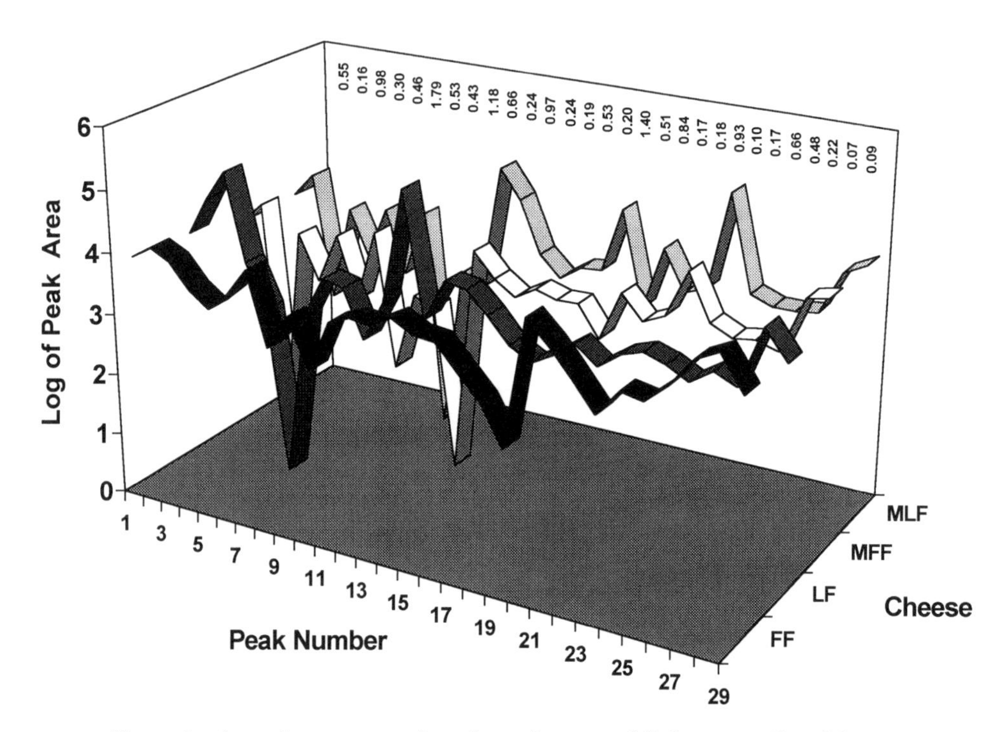

Figure 1. Area plot representative of two cheeses, a full-fat mature Scottish Cheddar and a mature reduced-fat cheese, by both conventional *in vitro* headspace analysis and *in vivo* analysis from mouth. FF = full-fat conventional; LF = reduced-fat conventional; MFF = full-fat, analysis from mouth; MLF = reduced-fat, analysis from mouth. Pooled SD values for all compounds, calculated from the replicate analyses, are given.

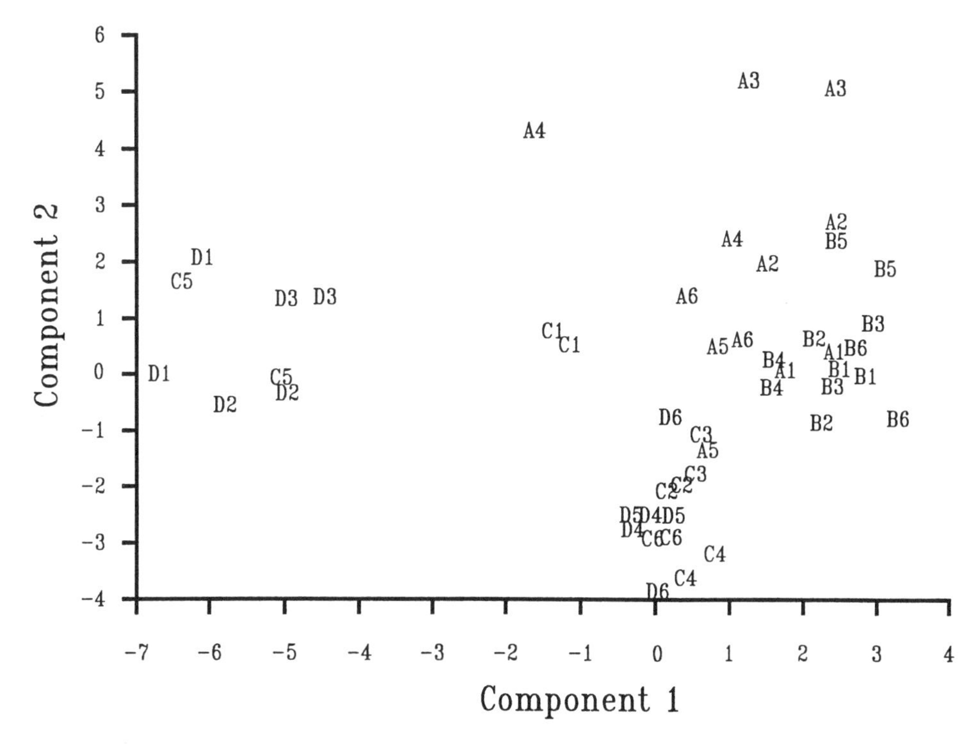

Figure 2. Principal component scores of the compounds identified in the chromatograms. Components 1 and 2 account for 24% and 14% of the variance respectively. A1–A6 are full-fat cheeses analyzed conventionally. C1–C6 are the same cheeses analyzed by buccal headspace methodology. B1–B6 and D1–D6 are reduced-fat cheeses analyzed by both methods as above.

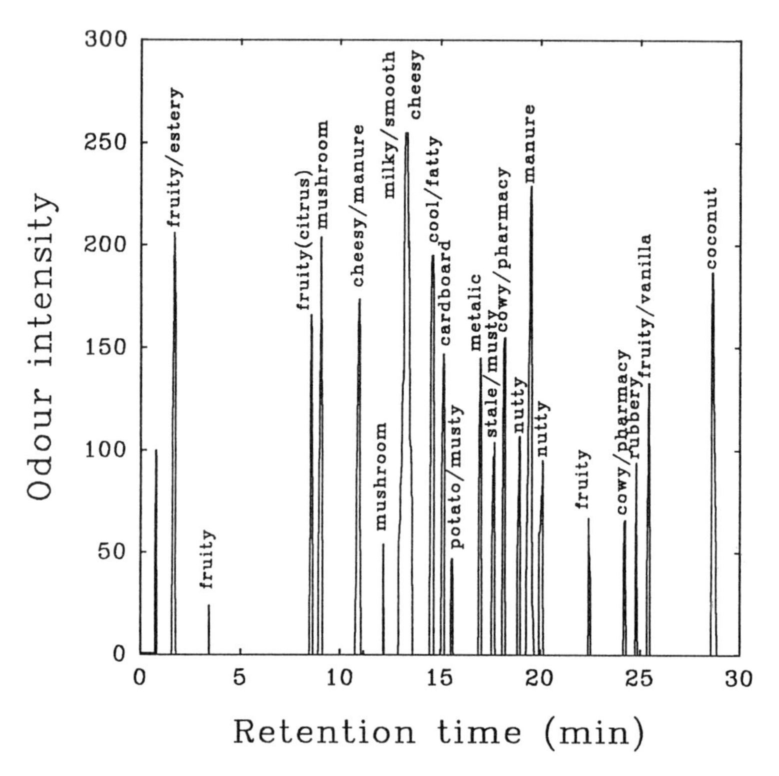

Figure 3. Aromagram of a mature Scottish Cheddar, conventional *in vitro* headspace analysis. Column details and temperature program are in the text.

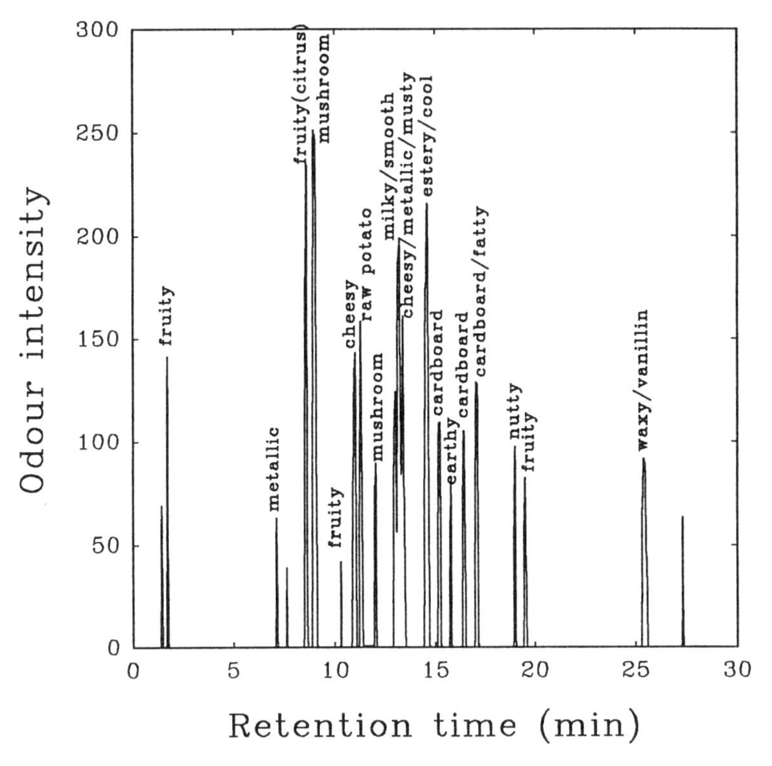

Figure 4. Aromagram of a mature Scottish Cheddar, *in vivo* headspace analysis. Column details and temperature program are in the text.

the *in vivo* method, although there were some exceptions. For example, there was a fruity note (retention time 8.67 min) which appeared to be of greater intensity in the *in vivo* extract, and a raw potato note (retention time 10.67 min) was found only in the *in vivo* extract. Those compounds associated with the most intense of the *in vivo* odors have been identified (Table I).

Table I. Intense Odor Notes Common to *in vivo* and *in vitro* Cheese Headspace Extract, and Most Intense in the *in vivo* Extract

Retention time (min)	*Odor quality*	*Compound Identified*
8.67	Fruity - intense	Hydroxy-heptanone[a]
10.67	Cheesy	Dimethyl trisulfide
13.00	Cheesy/metallic	3-(Methylthio)propanal
14.25	Cool/fatty	Dodecanal[a]

[a] Tentative identification by MS comparisons.

Sensory analysis. The Cheddar cheese vocabulary selected is shown in Table II. Loadings of descriptors on principal component 1 (which accounted for 53% of the total variance) from a PCA of the 8 mild cheeses are given in Table II, as are the ANOVA results on descriptors. Positive loadings were most associated with full-fat cheeses, negative loadings with reduced-fat cheeses. ANOVA results show the most discriminating descriptors.

Figures 5 and 6 show the averaged T-I curves for the descriptors cheesy and fruity. For the full-fat cheeses onset of the cheesy flavor was faster, more intense and remained longer than for the reduced-fat equivalent. On the other hand, onset of the fruity flavor was faster, more intense and persisted longer for the reduced-fat cheeses. Comparison of cheesiness with fruitiness reveals that maximum intensity for fruity was less than half that for cheesy, onset was slower and the flavor persisted longer. Curves varied considerably between assessors, but were reproducible for individual assessors. The same was found for swallow times.

Discussion

Many methods have been used to obtain "flavor extracts" from cheeses. The composition of these extracts varies both quantitatively and qualitatively depending on the method applied (*11*) often resulting in differences in opinion concerning the important cheese flavor compounds. For example, Vandeweghe and Reineccius (*12*) compared steam distillation, dialysis and solvent extraction on samples of Cheddar cheese. They found that the relative concentrations of the chemicals isolated varied, as did the aroma character of the isolate. Conventional methods of headspace analysis entail analysis of headspace volatiles above a food usually allowed to equilibrate in a

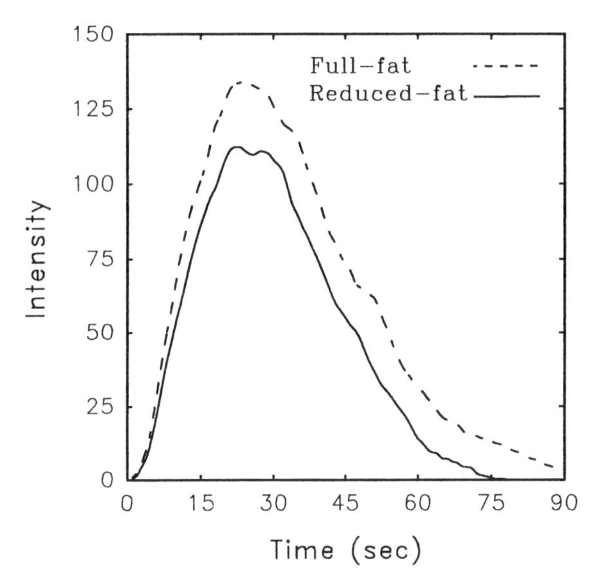

Figure 5. Mean time-intensity curve for perceived cheesiness in full-fat and reduced-fat Cheddar cheeses. Each curve represents the average of curves from ten panelists, replicated, for two cheeses of the same fat content.

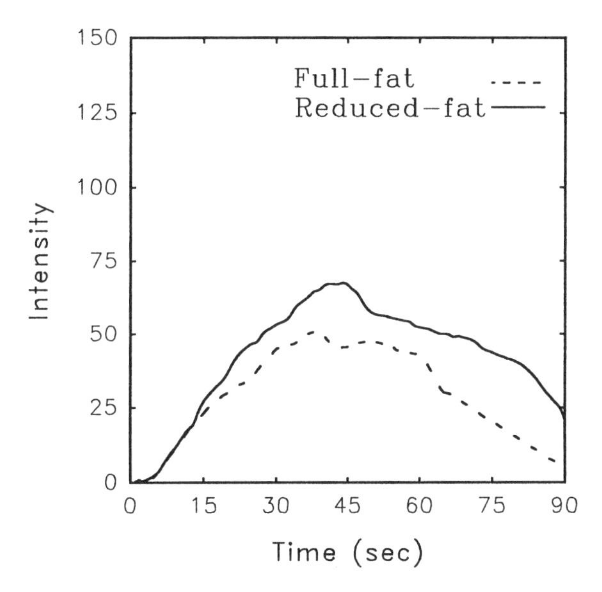

Figure 6. Mean time-intensity curve for perceived fruitiness in full-fat and reduced-fat Cheddar cheeses. Each curve represents the average of curves from ten panelists, replicated, for two cheeses of the same fat content.

sealed vessel. These do not account for the changes in structure and composition which a food may undergo upon mastication in the buccal cavity which affects the release of its volatile flavor components. Conventional methods, also, cannot account for losses of the released flavor which occur in the mouth and airways due to absorption (*13*). In view of this, measurement of volatile release in the mouth is attracting interest (*7, 14-16*).

Table II. Vocabulary Selected for Descriptive Analysis of Eight Mild Cheddar Cheeses

Descriptor	*Loading PC1*[a]	p^b
Cheesy	0.31	<0.001
Milky	-0.25	NS[c]
Buttery	0.02	NS
Salty	0.30	<0.05
Fruity	-0.06	NS
Moldy	0.29	NS
Rancid	0.27	<0.05
Sour	0.33	<0.001
Bitter	0.33	<0.01
Nutty	0.13	NS
Sulfurous	0.33	NS
Smokey	0.28	NS
Processed	-0.23	NS
Strength/Maturity	0.34	<0.001

[a] Loadings of descriptors on the first principal component. Positive loadings tend towards full-fat Cheddar cheeses, negative loadings towards reduced-fat Cheddar cheeses.
[b] Significance levels from an analysis of variance on descriptor for the Cheddar samples.
[c] NS = not significant.

Fat content, with which we were concerned, contributes to the perceived flavor of Cheddar in two ways. It is firstly a precursor of the taste and aroma compounds generated during production and maturation (the contribution of fat as a precursor to Cheddar flavor development is not thought to be of the greatest importance (*17-20*)), and secondly it acts as a medium that regulates the distribution of these and other compounds between water, fat and vapor phase (*21*). Most volatiles present in cheese are hydrophobic. As such they are present in the fat-phase of a cheese. Therefore, if the fat/water balance in a cheese is altered, there will be a resulting change in the release behavior of these volatiles. Addition of water, e.g. saliva, increases the rate of volatilization in most cases, because water is a poor solvent for these compounds. The more lipophilic is the compound, the greater is the effect (*22*).

From the results we obtained when comparing *in vitro* and *in vivo* headspace (Figures 1 and 2), it was quite clear that headspace obtained from the buccal cavity was quite distinct from that obtained *in vitro*. The conventional headspace analysis separated the full-fat and reduced-fat cheeses quite well. This separation was not as clear with headspace gathered from the mouth. Four cheeses were separated from the others, suggesting a flavor release difference not found by conventional means. Three of these were reduced-fat cheeses demonstrating the interaction between fat content and volatile release upon mastication in the mouth which was seen in the analyses of variance. Cheese C5, a full-fat cheese joins the reduced-fat. This cheese was also shown to be quite dissimilar from the other full-fat cheeses by conventional means (A5). Cheese C1 was also separated from the other full-fat cheeses, but to a lesser extent, as was the case for the conventional headspace data (A1).

Sniffport analysis was performed on a full-fat cheese. Conventional and buccal headspace methods were compared. As reported the *in vitro* extract was the more concentrated (Figures 3 and 4). This was as expected due to the disparity in sampling times, exhalation losses and tissue adsorption. Even so there were differences between the two not related to concentration. Relative odor intensities for some compounds were different (e.g. fruity, retention time 8.67 min), and odor notes not present *in vitro* were found in buccal headspace (e.g. raw potato, retention time 10.67 min). This supports flavor release behavior theory. The method of obtaining the *in vivo* headspace extract, where the cheese was consumed as it normally would be (see experimental), allows one to believe that compounds present in this extract must contribute to the flavor of the cheese. One in particular, 3-(methylthio)propanal (retention time 13.00 min), previously unreported in Cheddar flavor extracts, produced an intense cheesy/metallic/vegetable type odor in both headspace extracts. 2-Heptanone, associated with Cheddar flavor by some workers (*23, 24*) did not reach odor threshold in either extract. While sniffport methodology may be useful to identify important flavor compounds, no allowance is made for the effects of combinations of odors on perception. Sensory assessment of the entire cheese, in the form which it is normally consumed, can account for these problems but at the cost of increased complexity.

We have only used sensory descriptive analysis for vocabulary development, which can be used in further work, and to choose descriptors for the purposes of time-intensity scaling. This study revealed descriptors most associated with mild full-fat cheeses as: cheesy, salty, moldy, rancid, sour, bitter, nutty, sulphurous, smokey, and strength/maturity, and terms most associated with mild reduced-fat cheeses as: milky, fruity and processed. The term buttery could not be used to discriminate fat content. Individual cheeses could be discriminated using the terms: cheesy, salty, rancid, sour, bitter and strength/maturity. Because only eight mild cheeses were assessed (for time-intensity descriptor selection) and they were relatively similar (selected for homogeneity), other descriptors may have been simply redundant.

Temporal sensory assessment is useful for flavor evaluation in relation to cheese fat-content as the function of fat as a regulator of the distribution of the taste and aroma compounds of the cheese also influences the perception of these compounds by the sense organs. Volatile flavor compounds have a "threshold" concentration below which they effect little sensation, and above which their characteristic flavor is apparent. For example, a high fat content may eliminate the stimulus of a particular odor, or failing to do so, may reduce its intensity. This can be advantageous given that the perception of a particular odor may be quite different, and more unpleasant, at higher concentrations. Therefore a cheese which has a reduced fat content may have an earlier release of some aroma compounds as it is eaten, and the nature and time scale of the perceived sensation may be quite different, and probably less pleasant, than when a full-fat cheese is consumed (*21*). Temporal assessment of the flavor attributes cheesy and fruity (which might be considered an off-flavor (*25, 26*)) revealed differences in the perception of each attribute, and differences between cheeses according to fat content within each attribute. The balance of perceived flavor attributes influences overall flavor perception. For full-fat cheeses, maximum fruitiness was delayed by 20 sec and then perceived at a considerably lower intensity than its perception in the reduced-fat equivalent. This may be due to masking by the more intense cheesy flavor. Fruitiness also persists longer in the mouth after consumption. This may be due to adsorption in the buccal cavity tissues or to slower release behavior from the cheese.

The experiments outlined were not performed for in-depth analysis of cheese flavor, but to assess the possibility of applying this approach to cheese flavor evaluation. Results obtained indicate the potential of these methods. Unfortunately buccal headspace methodology in its present form is not ideal. Tenax traps are selective and some compounds which may have a key role to play in flavor are not trapped and therefore not analyzed. Non-volatile compounds, which may make a contribution to taste, suffer the same fate. Accordingly methods for measuring such compounds must be developed.

Conclusions

Reproducible cheese headspace data, displaced from the point of olfaction (the nose) have been obtained by the new buccal headspace methodology described. Differences were detected between full- and reduced-fat cheeses and between headspace samples taken *in vitro* and from the mouth. There was an interactive effect between the fat content of the cheese and the method of analysis. Sniffport analysis of buccal headspace revealed a number of compounds which are most likely to contribute to Cheddar flavor. Time-intensity scaling showed fat related differences between cheeses, and differences in the rates of perception of cheesy and fruity flavor. It is hoped that "buccal headspace" analysis will eventually help explain some of the loss in flavor quality associated with reduction of fat content of the cheeses. The methodology should be of great value in improving understanding of the dynamics of flavor release not just from cheese but from many foods, and in improving the accuracy of sensory correlation and consumer choice in relation to composition.

Acknowledgments

The authors wish to thank the UK Biotechnology and Biological Sciences Research Council (formerly the Agricultural and Food Research Council) and The Chivas and Glenlivet Group for financial support. C. M. Delahunty would like to thank the European Commission from whom he receives support.

Literature Cited

1. Maarse, H.; Visscher, C. A.; Willemsens, L. C.; Boelens, M. H. In *Volatile Compounds in Foods: Qualitative and Quantitative Data*; 6th edition; TNO-CIVO Food Analysis Institute: Zeist, The Netherlands, 1989.
2. von Sydow, E. *Food Tech.* **1971**, *25*, 40-45.
3. Lancet, D. In *Sensory Transduction*; Corey, D. P.; Roper, S. D., Eds.; Rockefeller University: New York, 1992, pp 73-91.
4. Piggott, J. R. *Food Qual. and Pref.* **1994**, *5*, 167-171.
5. Jameson, G. W. *Aust. J. Dairy Tech.* **1990**, *45*, 93-98.
6. Rosenberg, M. *Dairy Foods* **1992**, *93*, 44, 48.
7. Delahunty, C. M.; Piggott, J. R.; Conner, J. M.; Paterson, A. In *Trends in Flavour Research*; Maarse, H.; van der Heij, D.G. Eds.; Elsevier Applied Science: Amsterdam, 1994, pp. 47-52.
8. Piggott, J. R.; Sharman, K. In *Statistical Procedures in Food Research*; Piggott, J. R. Ed.; Elsevier Applied Science: London, 1986, pp 181-232.
9. Piggott, J. R.; Mowat, R. G. *J. Sens. Stud.* **1991**, *6*, 49-62.
10. Oreskovich, D.C.; Klein, B.P.; Sutherland, J.W. In *Sensory Science Theory and Applications in Foods;* Lawless, H.; Klein, B. P. Eds.; Marcel Dekker: New York, 1991, pp 353-393.
11. Springett, M. *Food Rev.* **1991**, *18*, 16-18.
12. Vandeweghe, P.; Reineccius, G. A. *J. Agric. Food Chem.* **1990**, *38*, 1549-1552.
13. Overbosch, P.; Afterof, W. G. M.; Haring, P. G. M. *Food Rev. Int.* **1991**, *7*, 137-184.
14. Soeting, W. J.; Heidema, J. *Chem. Sens.* **1988**, *13*, 607-617.
15. Linforth, R. S. T.; Taylor, A. J. *Food Chem.* **1993**, *48*, 115-120.
16. Taylor, A. J.; Linforth, R. S. T. In *Trends in Flavour Research*; Maarse, H.; van der Heij, D. G. Eds.; Elsevier Applied Science: Amsterdam, 1994, pp 3-14.
17. Aston, H. W.; Dulley, J. R. *Aust. J. Dairy Tech.* **1982**, *37*, 59-64.
18. Manning, D. J.; Price, J. C. *J. Dairy Res.* **1977**, *44*, 357-361.
19. Law, B. A. In *Advances in the Microbiology and Biochemistry of Cheese and Fermented Milk*; Davies, F. L.; Law, B. A. Eds.; Elsevier Applied Science: London, 1984, pp 187-208.
20. Wong, N. P.; Ellis, R.; La Croix, D. E. *J. Dairy Sci.* **1975**, *58*, 1437-1441.

21. Gurr, M. I.; Walstra, P. *Bull. Int. Dairy Fed.* **1989**, *244*, 44-46.
22. McNulty, P. B.; Karel, M. J. *J. Food Tech.* **1973**, *8*, 309-318.
23. Yoshizawa, T.; Yamamoto, I.; Yamamoto, R. *Botu-Kagaku-Sci. Pest Cont.* **1970**, *35*(II), 43-45.
24. Walker, N. J.; Keen, A. R. *J. Dairy Res.* **1974**, *41*, 73-80.
25. Aston, J. W.; Gilles, J. E.; Durward, I. G.; Dulley, J. R. *J. Dairy Res.* **1985**, *52*, 565-572.
26. Urbach, G. (1993). *Int. Dairy J.* **1993**, *3*, 389-422.

Chapter 19

Flavoring of Complex Media: A Model Cheese Example

C. Dubois[1], M. Sergent[2], and A. Voilley[1]

[1]Ensbana, 1 Esplanade Erasme, 21000 Dijon, France
[2]LPRAI, 13397 Marseille Cedex 13, France

This study deals with the flavoring of a fresh analog cheese. The effect of cheese composition on structural characteristics (the cheese hardness and fat distribution) has been investigated, as well as the physico-chemical interactions between the cheese constituents and two cheese aroma compounds. Diacetyl and diallyl sulfide were chosen to represent cheese flavorants, which depend on the volatilities of a diversity of aroma compounds. Their vapor-lipid partition coefficients were measured via a headspace technique. An experimental design was used to show that the cheese composition influenced its hardness. Volatility differences as a function of varying cheese composition depended on the nature of the aroma compound, but seemed to be independent of the fat distribution within the cheese.

In response to consumers' wishes, the food industry is constantly introducing novel formulations which contain new ingredients or aroma compounds. Because foods are complex systems, the field for formulation is very broad. The use of experimental design thus offers a great advantage by minimizing the number of experiments necessary to study of the effect of formulation changes on the organoleptic qualities of the product (*1*). Otherwise, an in-depth knowledge of the behavior of aroma compounds in food is necessary to select aromatic raw materials and optimize their use in a given process or product.

Literature research has used sensory evaluation or instrumental measurements to gain a better understanding of the mechanisms that occur between aroma compounds and non-volatile substances. The systems considered are often very simple, consisting of an aroma compound and a single constituent, usually in an aqueous solution. In general, the presence of proteins (*2, 3*), polysaccharides (*4, 5*), lipids (*6, 7*) or trace lipids (*8*) reduces the volatility of aroma compounds with

0097–6156/96/0633–0217$15.00/0

respect to that in pure water. On the other hand, the presence of salts increases their volatilities. Recent papers have also been published concerning studies on flavor release in the mouth (*9, 10*). This chapter concerns the flavoring of a model cheese. The use of a model cheese allowed reproducible formulations in the laboratory and the ability to change composition easily. An experimental design of mixtures allowed the effect of the ingredients on the cheese hardness and the distribution of the fat content to be studied, as well as the interactions with aroma compounds. Diacetyl was considered to be an example of an aroma compound produced by lactic bacteria, while diallyl sulfide is a compound typically added to garlic-flavored cheeses.

Materials and Methods

Aroma Compounds. The two compounds chosen differed in their chemical functional groups and hydrophobicity constants (Table I).

Table I. Physico-chemical and Sensory Characteristics of Aromas

Compound	*Formula*	*MW*	*Solubility in water (g/L)*	*LogP*[a]	*Odor*
Diacetyl	CH_3-CO-CO-CH_3	86	250 (15 °C)	- 2.00	butter
Diallyl sulfide	CH_2=CH–CH_2 –S–CH_2=CH–CH_2	114	0.99 (25 °C)	+ 2.42	garlic, pungent

[a] Hydrophobicity constant calculated according to Rekker's method (*16*)

The Model Cheese.

Ingredients. The cheese analog was composed of distilled water, calcium caseinate (Unilait, France), sodium chloride and anhydrous milk fat (Fracexpa, France). The pH was adjusted to 4.9 with lactic acid (Prolabo, France; purity 99%).

Processing. The model cheese was made from a heated oil-in-water emulsion stabilized by calcium caseinate. Gelification was achieved by addition of lactic acid. Mixing was carried out with a Kenwood kitchen mixer.

The Cheese Flavoring. Diallyl sulfide and diacetyl were introduced into the cheese at a concentration of 100 ppm (v/w); the flavored cheeses were stored at room temperature for 3 days before analysis.

Composition of Cheeses Studied. The three parameters varied in the experimental design were the content of water, fat and calcium caseinate as follows:

$$57 \leq \text{Water content} \leq 76$$

$$11 \leq \text{Calcium caseinate content} \leq 23$$

$$0 \leq \text{Fat content} \leq 30$$

The sodium chloride content was the same in all cheeses (NaCl = 1.4%). The quantity of lactic acid added to maintain a constant pH (4.9) depended on the calcium caseinate content. An experimental design of mixtures with constraints was adopted; the limits were defined to cover a range of ratios of fat to dry matter between 0-70%. Defining two out of three contents sufficed, as the third content was the complement to 100%.

The defined experimental domain is shown in Figure 1 with the nine experimental points needed to model the responses. The analog cheeses were made at three different levels of calcium caseinate, three water levels, and nine fat contents. The results were processed with Nemrod software (D. Mathieu and R. Phan Tan Luu, LPRAI, Marseille, France) to obtain a multiple linear regression of the three-dimensional curve representing the response (hardness, volatilities) versus two independent composition variables. The validity of the model was checked according to the standard deviations of the experimental results (around 15% for the texture measurements, less than 8% for both compounds' volatilities and the interfacial area calculations).

Studied Responses.

Texture, expressed as hardness, was evaluated at 25 °C by penetrometry with a Stevens apparatus. The maximum force at a 1-cm penetration with a 3-mm diameter probe was recorded.

Fat Distribution in the Core of the Cheese was determined by laser granulometry (Malvern Mastersizer S2-01, Malvern Instrument, Worcs, U.K.). The specific interfacial area (m^2/100 g of cheese) was calculated.

Volatilities of Aroma Compounds in the Cheese were determined by equilibrium headspace analysis. Interactions between aroma compounds and non-volatile constituents may be of two types: reversible and low energy or irreversible and high energy.

In the latter case, the sensorial perception of the aroma compounds is qualitatively changed; in our study, only reversible physico-chemical interactions of systems at equilibrium were considered. Interactions occur at the molecular level and are expressed at a macrocospic level as changes in the equilibrium between the product and the vapor phase, and between immiscible phases in the core of the cheese: the vapor-liquid partition coefficient, *k*, was determined by headspace analysis (*11*).

$$k = \frac{\textit{Aroma concentration in the vapor phase}\ (g\,/\,L)}{\textit{Aroma concentration in the cheese}\ (g\,/\,kg)}$$

Results

Effect of Composition on Hardness. The hardness of the model systems in the experimental domain is shown in Figure 2. At constant caseinate content, the water and fat contents changed in opposite directions so that the sum of the three constituents was 1. The variations in the hardness were very large as some products were fluid with no rigidity (cheeses no. 3, 6, 8), while cheese no. 2 was very

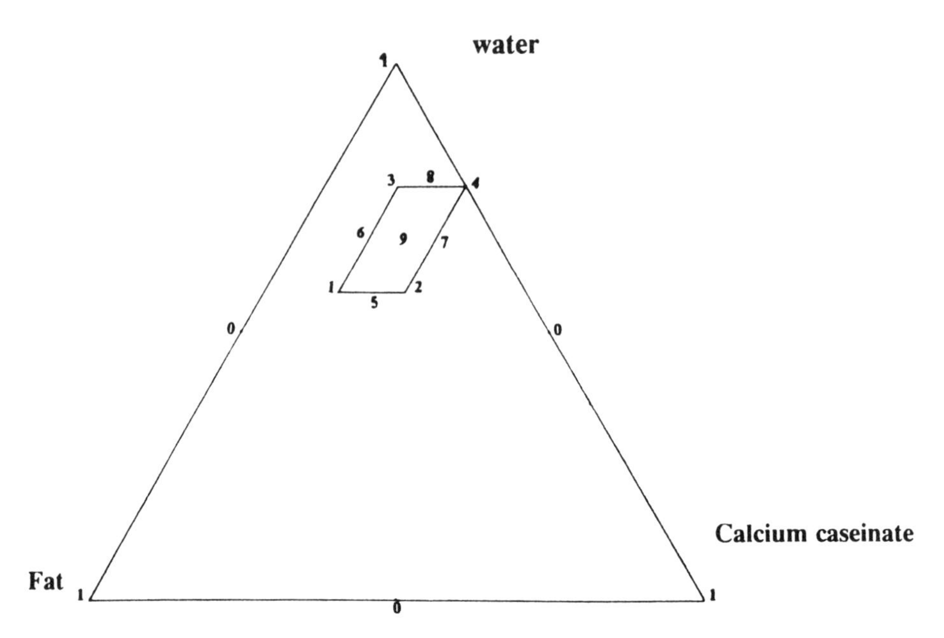

Figure 1. Representation of the experimental domain.

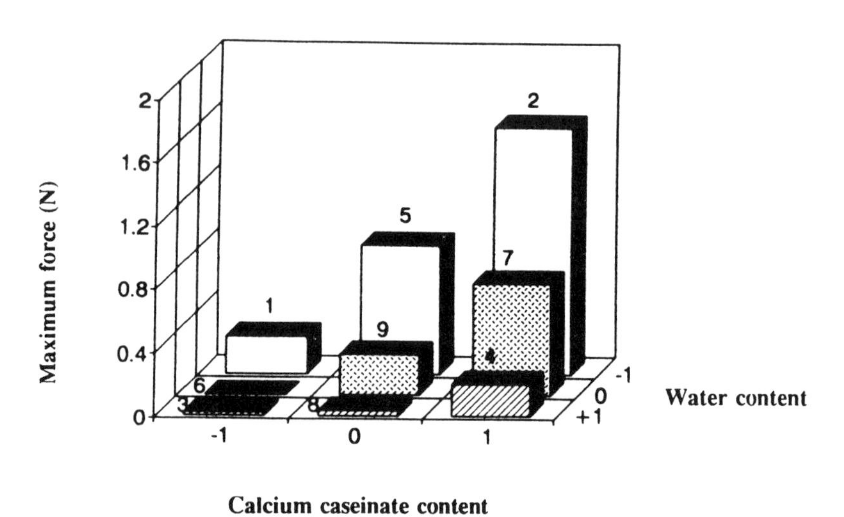

Figure 2. Maximal force, *N*, as a function of cheese composition (no. 1–9).

compact and crumbly. The hardness was affected mainly by the water and calcium caseinate contents and very little by that of fat. At constant calcium caseinate content, the reduction of the water content conferred a solid consistency to products which were fluid or induced significant hardening in products already solid. An increase in protein content resulted in strengthening of the network due to calcium ion binding.

Effect of Composition on Fat Distribution. Cheese can be considered to be an emulsion or an emulsified gel stabilized by caseinate. The distribution of fat in the cheese and the water interfacial area varied with the composition (Figure 3). The specific interfacial area increased with the fat content as expected. Additionally, for cheeses no. 7 and 3, and cheeses no. 6 and 2, a significant increase in specific interfacial area with increasing calcium caseinate content was observed. The fat distribution in the product can be an important factor as far as interactions with aroma compounds are concerned; it should affect the sensory perception, notably because of the change of diffusion parameters in the mouth.

Effect of Composition on the Volatility of Aroma Compounds.

Diallyl Sulfide. Results are shown in Figure 4. The volatility of diallyl sulfide was affected principally by the fat content. Volatility decreased sharply up to 15% fat and changed very little thereafter. This type of exponential decrease has been observed previously by Buttery *et al.* (*12*) in biphasic, non emulsified water-lipid systems. The calcium caseinate content did not appear to be an important factor in that case, but did affect the interfacial area. For cheeses no. 7 and 3, as well as for cheeses no. 6 and 2, the volatility decreased slightly when the surface area (Figure 3) increased.

Diacetyl. The volatility of diacetyl in the different cheeses is shown in Figure 5. Within the experimental domain, the variation was less than for diallyl sulfide. However, a lowering of k with increasing caseinate content could be observed. At constant caseinate content, the variation of the volatility when the water content increased (i.e., when the fat content decreased) was very small; a change in the fat content of up to 30% did not affect the volatility.

Effect of Fat Type on the Volatility of Aroma Compounds. Milk fat, which was partly solid at room temperature, was replaced by tributyrin (Aldrich, France, purity 98%), which was liquid at 25°C. Volatilities of diacetyl and diallyl sulfide are given in Table II. The volatility of diacetyl was not affected by the nature of the fat. In contrast, the volatility of diallyl sulfide decreased by 20% in the presence of tributyrin. The importance of the nature of the fat phase was thus demonstrated as regards the product formulation.

Discussion

Differences in the volatility of aroma compounds in cheeses with different compositions have been shown. In order to better understand these phenomena, physico-chemical interactions between aroma compounds and cheese constituents

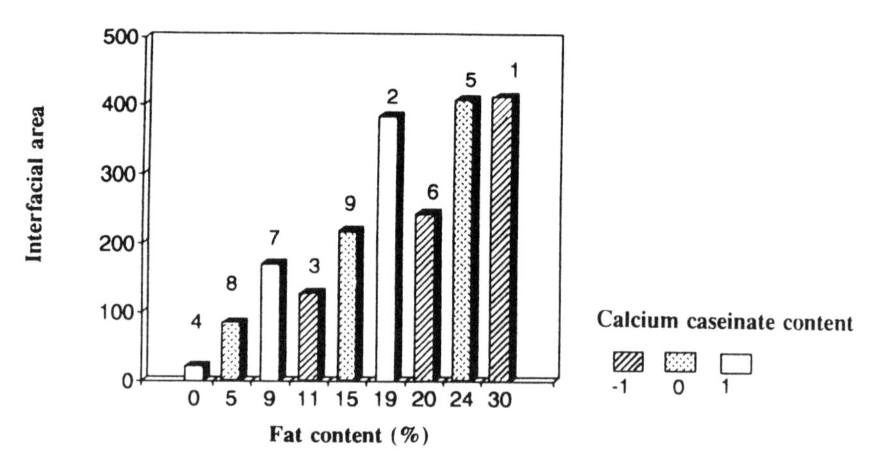

Figure 3. Specific interfacial area (m^2/100 g of cheese no. 1–9) vs. cheese composition.

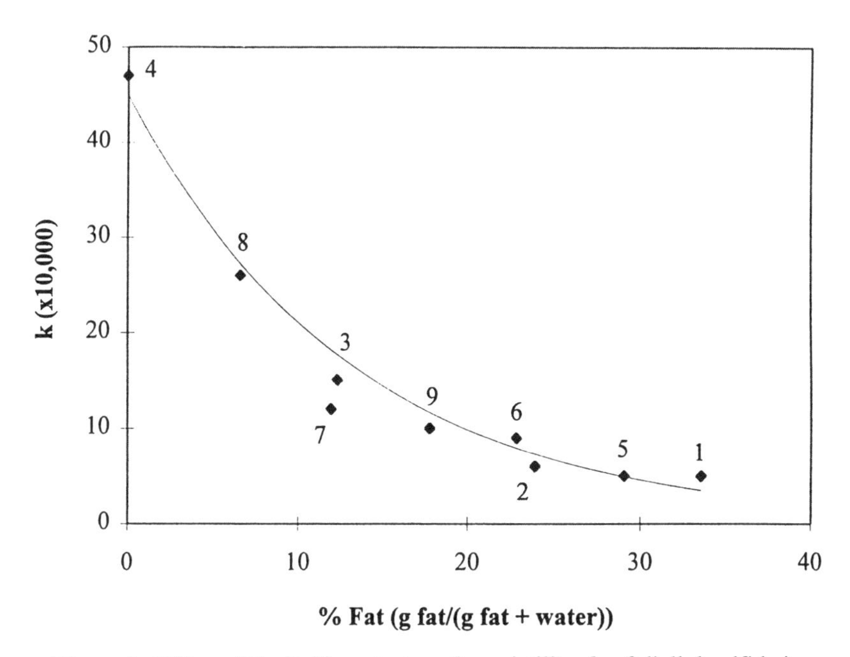

Figure 4. Effect of the lipid content on the volatility, k, of diallyl sulfide in model cheeses (no. 1–9) at 25 °C and 760 mm Hg.

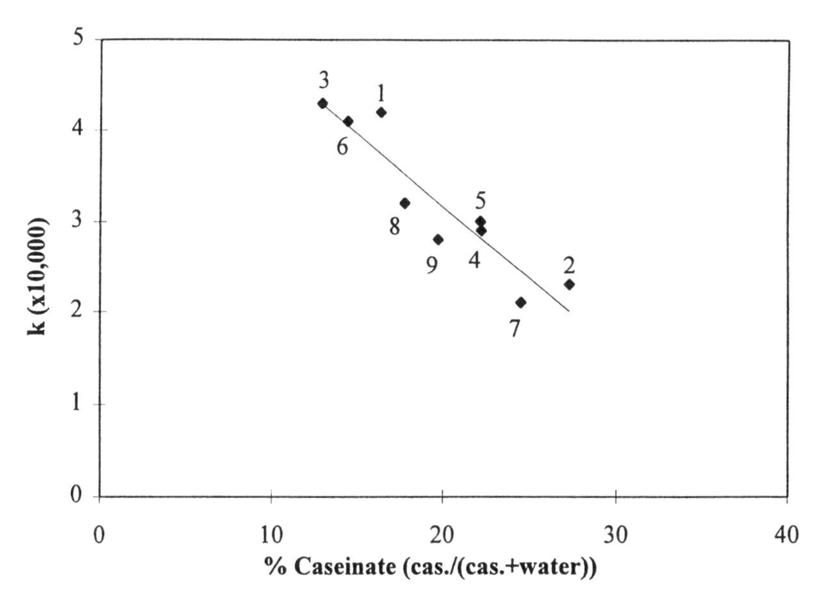

Figure 5. Volatility, k, of diacetyl in cheeses (no. 1–9) at 25 °C and 760 mm Hg.

have been considered. The vapor-liquid partition coefficients, *k*, of both aroma compounds in the water or lipid phases are listed in Table III.

Table II. Volatilities of Diacetyl and Diallyl Sulfide in Cheese as a Function of Lipid Type

Lipid type	*Diacetyl* $k\ (x\ 10^4)$	*Diallyl sulfide* $k\ (x\ 10^4)$
AMF[a]	2.51	8.60
Tributyrin	2.59	6.77

[a] AMF: Anhydrous milkfat

Table III. Vapor-Liquid Partition Coefficients of Aroma Compounds in Water and in Lipid Phases (25 °C)

Medium	*Diacetyl* $k\ (x\ 10^4)$	*Diallyl sulfide* $k\ (x\ 10^4)$
Water	5.0	713
AMF[a]	14.3	2.1
Tributyrin	5.0	1.0

[a] AMF: Anhydrous milkfat

Diallyl sulfide (a hydrophobic molecule) volatilized more readily than diacetyl from water, because diacetyl has a greater affinity for water. The opposite was observed in lipid media. The more hydrophobic the aroma compound, the larger its affinity for lipids. The influence of the lipid phase on the volatility, i.e., on physico-chemical interactions, could be attributed either to the nature of the lipid, especially due to its hydrophobicity, or to its physical state. For instance, Maier (*13*) showed that the sorption of aroma compounds to liquid triglycerides was greater than that to solid triglycerides. In our study, milk fat had 15% triglycerides in the solid state at 25 °C. Its overall hydrophobicity, arising from fatty acids with 4 to 18 carbon atoms, was greater than that of tributyrin, resulting in higher vapor-liquid partition coefficients in the AMF. Both the physical state and the nature of the lipid could indeed modify the volatility. A better flavor stability could thus be achieved by making such cheeses with liquid triglycerides – such as tributyrin – for which the vapor-liquid partition coefficient was the lowest observed (Tables II and III).

In a biphasic system comprised of two immiscible phases, the aroma compounds distribute themselves according to their affinities for each phase. The liquid-liquid partition coefficient, *P*, is the ratio of the aroma concentration in the lipid phase to the one in the aqueous phase, at equilibrium. Diallyl sulfide was concentrated much more in the lipid phase than diacetyl. At 25 °C, the liquid-liquid partition coefficients for allyl sulfide were 307 and 610, respectively, for water–tributyrin and water–anhydrous milkfat systems, while for diacetyl, *P*-values were 0.4 and 1.3, respectively, for water–tributyrin and water–anhydrous milkfat systems (*11*). The behavior of diacetyl is opposite to that of diallyl sulfide: it indeed had little affinity for fat and was little impacted in our domain of formulation, although its volatility was slightly reduced with caseinate (Figure 5), as already observed by Fares (14) and Dumont and Land (15) with diacetyl–protein binding. In any case, the butter-like flavor should be preserved in such high water content cheeses. Because diallyl sulfide had a high *P*-value, it was greatly concentrated in the fat phase as the cheese fat content increased; because the *k*-value in this phase was very low, the volatility of diallyl sulfide in the cheese decreased very rapidly in the presence of lipids and especially with tributyrin. For best flavoring results, in our experimental domain, the highest tributyrin percentage should be considered. It must be pointed out that this choice should only be made after sensory testing, since all flavoring compounds have their own sensory thresholds. The cheese hardness, which mainly depended on the water and caseinate contents, may be optimized for medium rigidity depending on the texture preferred by the consumer, for cheeses no. 5 and 9 with 57 to 66% water and approximately 17% caseinate and 20% fat.

Conclusion

The poor success of some low-calorie products, notably those low in fat, can sometimes be explained by a small change in the formulation leading to large changes in the behavior of aroma compounds. The distribution of aroma volatiles between water and lipid phases is modified, based on their physico-chemical properties, by other constituents. If the interactions between aroma compounds and food constituents were known, it would be possible to predict the effect of a change of the composition of complex food products. Consequently, complementary work is needed to better assess the effect of the liquid-liquid and vapor-liquid interfaces on the flavoring behavior of aroma compounds in such products.

Acknowledgments

The authors wish to thank Mr. Stéphane Fayoux for helping with the translation and the wording of the publication.

Literature Cited

1. Fargin, E.; Sergent, M.; Mathieu, D.; Phan Tan Luu, R. *Biosciences.* **1985**, *4*, 77-88.
2. Kinsella, J. E. *Int. New Fat Oils Relat. Mater.* **1990**, *1*, 215-226.
3. Hansen, A. P.; Heinis, J. J. *J. Dairy Sci.* **1992**, *75*, 1211-1215.
4. Sorrentino, F.; Voilley, A.; Richon, D. *AIChE J.* **1986**, *32*, 1988-1993.
5. Goshall, M. A.; Solms, J. *J. Food Technol.* **1992**, *46*, 140-145.

6. Ebeler, S. E.; Pangborn, R. M.; Jennings, W. G. *J. Agric. Food Chem.* **1988**, *36*, 791-796.
7. Le Thanh, M.; Voilley, A.; Phan Tan Luu, R. *Sci. des Aliments.* **1993**, *13*, 699-710.
8. Lubbers, S.; Charpentier, C.; Feuillat, M.; Voilley, A. *Am. J. of Enology and Viticulture* **1993**, *45*, 29-33.
9. De Roos, K. B.; Wolswinkel, K. In *Trends in Flavour Research. Proceedings of the 7th Weurman Flavour Research Symposium*, Noordwijkerhout, The Netherlands, 15-18 June 1994; Maarse H.; Van Der Eij D. J. Eds.; Elsevier: London, 1994; pp 15-32.
10. Delahunty, C. M.; Piggot, J. R.; Cooner, J. M.; Paterson, A. In *Trends in Flavour Research. Proceedings of the 7th Weurman Flavour Research Symposium*, Noordwijkerhout, The Netherlands, 15-18 June 1994; Maarse H.; Van Der Eij, D. J., Eds.; Elsevier: London, 1994; pp 47-52.
11. Barbier-Dubois, C. PhD Thesis, ENSBANA, Université de Bourgogne, Dijon, France, 1994.
12. Buttery, R. G.; Guardini, D. G.; Ling L. C. *J. Agric. Food Chem.* **1973**, *21*, 198-201.
13. Maier, H. G. *Proc. Int. Symp. Aroma Research*, Zeist, Wageningen, **1975**, 143-157.
14. Dumont, J. P.; Land, D. G. *J. Agric. Food Chem.* **1986**, *34*, 1041-1045.
15. Fares, K. PhD Thesis, ENSBANA, Université de Bourgogne, Dijon, France, 1987.
16. Rekker, R. *The Hydrophobic Fragmental Constant*, Nauta W.; Rekker, R., Eds.; Elsevier Scientific Publishing Company: Amsterdam, 1977.

Author Index

Affiliation Index

Subject Index